P9-CFO-319

# HIGGS

# HIGGS

*The Invention and Discovery of the 'God Particle'*

JIM BAGGOTT

OXFORD
UNIVERSITY PRESS

**OXFORD**
**UNIVERSITY PRESS**

Great Clarendon Street, Oxford, OX2 6DP,
United Kingdom

Oxford University Press is a department of the University of Oxford.
It furthers the University's objective of excellence in research, scholarship,
and education by publishing worldwide. Oxford is a registered trade mark of
Oxford University Press in the UK and in certain other countries



First Edition published in 2012

Impression: 1



British Library Cataloguing in Publication Data
Data available

Library of Congress Cataloging in Publication Data
Data available

ISBN 978–0–19–960349–7

Printed in Great Britain by
Clays Ltd, St Ives plc

For Ange

# CONTENTS

# ABOUT THE AUTHOR

Jim Baggott is an award-winning science writer. A former academic scientist, he now works as an independent business consultant but maintains a broad interest in science, philosophy, and history, and continues to write on these subjects in his spare time. His previous books have been widely acclaimed and include:

*The Quantum Story: A History in 40 Moments* (Oxford University Press, 2011);

*Atomic: The First War of Physics and the Secret History of the Atom Bomb 1939–49* (Icon Books, 2009), shortlisted for the Duke of Westminster Medal for Military Literature, 2010;

*A Beginner's Guide to Reality* (Penguin, 2005);

*Beyond Measure: Modern Physics, Philosophy and the Meaning of Quantum Theory* (Oxford University Press, 2004);

*Perfect Symmetry: The Accidental Discovery of Buckminsterfullerene* (Oxford University Press, 1994); and

*The Meaning of Quantum Theory: A Guide for Students of Chemistry and Physics* (Oxford University Press, 1992).

# PREFACE

The news that something very much like the Higgs boson had been discovered, at CERN in Geneva on 4 July 2012, flashed instantaneously around the world like some highly contagious electronic virus. Headlines screamed of this latest triumph of high-energy physics. The discovery made front-page news, was featured in many evening news bulletins, and reached an audience of billions. Signals consistent with a particle that had first been hypothesized or 'invented' in 1964 had at last been found, 48 years later, at a cost of many billions of dollars.

So, what was all the fuss about? What is the Higgs boson and why does it matter so much? If this new particle really is the Higgs, what does it tell us about the material world and the evolution of the early universe? Was finding it really worth all the effort?

The answers to these questions can be found in the story of the so-called Standard Model of particle physics. As the name implies, this is the framework that physicists use to interpret the elementary constituents of all matter and the forces that bind matter together, or cause it to fall apart. It is a body of work built up over many decades of unstinting effort, which represents the physicists' best efforts to interpret the physical world around us.

The Standard Model is not yet a 'theory of everything'. It does not account for the force of gravity. In recent years you may have read about exotic new theories of physics which attempt to unify the fundamental forces, including gravity. These are theories such as supersymmetry and superstrings. Despite the efforts of hundreds of theorists engaged on these projects, these new theories remain speculative and have little or no supporting evidence from experiment. For the time being, and despite flaws that have been acknowledged since its inception in the late 1970s, the Standard Model is still where most of the real action is.

The Higgs boson is important in the Standard Model because it implies the existence of a Higgs field, an otherwise invisible field of energy which pervades the entire universe. Without the Higgs field, the elementary particles that make up you, me, and the visible universe would have no mass. Without the Higgs field mass could not be constructed and nothing could *be*.

It seems we owe quite a lot to the existence of this field. This is one of the reasons why the Higgs boson, the particle of the Higgs field, has been hyped in the popular press as the *God particle*. This is a name heartily despised by practising scientists, as it overstates the importance of the particle and draws attention to the sometimes uneasy relationship between physics and theology. It is, however, a name much beloved by science journalists and popular science writers.

Many of the predicted consequences of the Higgs field were borne out in particle collider experiments in the early 1980s. But inferring the field is not the same as detecting its tell-tale field particle. It is therefore immensely reassuring to know that the field is very probably here, there, and everywhere. The possibility that the Higgs boson might not have been found was very real, and the implications for the Standard Model were potentially devastating.

I began writing this book in June 2010, two years before the discovery was made. I had just completed the manuscript of another book, called *The Quantum Story: A History in 40 Moments* which, as the title implies, is a history of quantum physics from 1900 to the present day. That book covered the development of the Standard Model and the invention of the Higgs field and its particle. A few months earlier CERN's Large Hadron Collider reached record proton–proton collision energies of seven trillion electron volts, and I figured that a discovery *might* be possible within the next few years. Happily, I was proved right.

*The Quantum Story* was published in February 2011. The present book is based, in part, on that earlier work.

My thanks go to Latha Menon and the delegates at Oxford University Press, who were ready to risk commissioning a book about a particle that hadn't yet been discovered. I have followed developments at CERN through the official channels, but acknowledge a debt to a number of high-energy physics bloggers, including Philip Gibbs, Tommaso Dorigo, Peter Woit, Adam Falkowski, Matt Strassler and Jon Butterworth. Thanks also to Jon Butterworth, Sophie Tesauri, James Gillies, Laurette Ponce, and Lyndon Evans for taking the time to talk to me and share their growing sense of excitement. I would also like to express my gratitude to Professors David Miller, and Peter Woit, who read and commented on the draft manuscript, and to Professor Steven Weinberg, who also read through the draft manuscript and kindly contributed a personal perspective in his Foreword. Be assured that the errors that remain are all my own work.

Jim Baggott
Reading, 6 July 2012.

# FOREWORD
## *by Steven Weinberg*

Many important scientific discoveries have been followed by popular books explaining these discoveries to general readers. But this is the first case I have seen of a book that has been largely written in *anticipation* of a discovery. The readiness of this book for publication immediately after the announcement in July 2012 of the discovery at CERN (with some corroboration from Fermilab) of a new particle that seems to be the Higgs particle testifies to the remarkable energy and enterprise of Jim Baggott and Oxford University Press.

The prompt publication of this book also testifies to the widespread public interest in this discovery. So it may be worthwhile if in this Foreword I add some remarks of my own about just what has been accomplished. It is often said that what was at stake in the search for the Higgs particle was the origin of mass. True enough, but this explanation needs some sharpening.

By the 1980s we had a good comprehensive theory of all observed elementary particles and the forces (other than gravitation) that they exert on one another. One of the essential elements of this theory is a symmetry, like a family relationship, between two of these forces, the electromagnetic force and the weak nuclear force. Electromagnetism is responsible for light; the weak nuclear force allows particles inside atomic nuclei to change their identity in radioactive decay processes.

This symmetry brings the two forces together in a single 'electroweak' structure. The general features of the electroweak theory have been well tested; their validity is not what has been at stake in the recent experiments at CERN and Fermilab, and would not be seriously in doubt even if no Higgs particle had been discovered.

But one of the consequences of the electroweak symmetry is that, if nothing new is added to the theory, all elementary particles including electrons and quarks would be massless, which of course they are not. So, something has to be added to the electroweak theory, some new kind of matter or field, not yet observed in nature or in our laboratories. The search for the Higgs particle has been a search for the answer to the question: What is this new stuff we need?

The search for this new stuff has not been just a matter of noodling around at high energy accelerators, waiting to see what turns up. Somehow the electroweak symmetry, an exact property of the underlying equations of elementary particle physics, must be broken; it must not apply directly to the particles and forces we actually observe. It has been known since the work of Yoichiro Nambu and Jeffrey Goldstone in 1960–61 that symmetry-breaking of this sort is possible in various theories, but it had seemed that it would necessarily entail new massless particles, which were known not to exist.

It was the independent work of Robert Brout and François Englert; Peter Higgs; and Gerald Guralnik, Carl Hagen and Tom Kibble, all in 1964, that showed that in some kinds of theories these massless Nambu-Goldstone particles would disappear, serving only to give mass to force-carrying particles.* This is what happens in the theory of weak and electromagnetic forces proposed in 1967–68 by Abdus Salam and myself. But this still left open the question: What sort of new matter or field is actually breaking the electroweak symmetry?

There were two possibilities. One possibility was that there are hitherto unobserved fields that pervade empty space, and that just as the earth's magnetic field distinguishes north from other directions, these new fields distinguish weak from electromagnetic forces, giving mass to the particles that carry the weak force and to other particles, but leaving photons (which carry the electromagnetic force) with zero mass. These are called 'scalar' fields, meaning that unlike magnetic fields they do not distinguish directions in ordinary space. Scalar fields of this general sort were introduced in the illustrative examples of symmetry-breaking used by Goldstone and later in the 1964 papers.

When Salam and I used this sort of symmetry-breaking in developing the modern 'electroweak' theory of weak and electromagnetic forces, we assumed that the symmetry breaking was due to fields of this scalar type, pervading all space. (A symmetry of this sort had already been hypothesized by Sheldon Glashow and by Salam and John Ward, but not as an exact property of the equations of the theory, so these theorists were not led to introduce scalar fields.)

One of the consequences of theories in which symmetries are broken by scalar fields, including the models considered by Goldstone and the 1964 papers and the electroweak theory

* For brevity, I will refer to this work as 'the 1964 papers'.

of Salam and me, is that although some of these fields serve only to give mass to the force carrying particles, other scalar fields would be manifested in nature as new physical particles that could be created and observed in accelerators and particle colliders. Salam and I found we needed to put four scalar fields into our electroweak theory. Three of these scalar fields were used up in giving mass to the $W^+$, $W^-$, and $Z^0$ particles – the 'heavy photons' – that in our theory carry the weak force (these particles were discovered at CERN in 1983–84, and found to have the masses predicted by the electroweak theory). One of the scalar fields was left over to be manifested as a physical particle, a bundle of the energy and momentum of this field. This is the 'Higgs particle' for which physicists have been searching for nearly thirty years.

But there was always a second possibility. There might instead be no new scalar fields pervading space, and no Higgs particle. Instead, the electroweak symmetry might be broken by strong forces, known as 'technicolour forces', acting on a new class of particles too heavy to have been observed yet. Something like this happens in superconductivity. This kind of theory of elementary particles was proposed in the late 1970s independently by Leonard Susskind and myself, and would lead to a whole forest of new particles, held together by technicolour forces. So this is the alternative with which we have been faced: Scalar fields? Or technicolour?

The discovery of the new particle casts a very strong vote in favour of the electroweak symmetry being broken by scalar fields, rather than by technicolour forces. This is why the discovery is important.

But much remains to be done to pin this down. The electroweak theory of 1967–68 predicted all of the properties of the Higgs particle, except its mass. With the mass now known experimentally, we can calculate the probabilities for all the various ways that Higgs particles can decay, and see if these predictions are borne out by further experiment. This will take a while.

The discovery of a new particle that appears to be the Higgs also leaves theorists with a difficult task, to understand its mass. The Higgs is the one elementary particle whose mass does not arise from the breakdown of the electroweak symmetry. As far as the underlying principles of the electroweak theory are concerned, the Higgs mass could have any value. That is why neither Salam nor I could predict it.

In fact, there is something puzzling about the Higgs mass we now do observe. It is generally known as the 'hierarchy problem'. Since it is the Higgs mass that sets the scale for the masses of all other known elementary particles, one might guess that it should be similar to another mass that plays a fundamental role in physics, the so-called Planck mass, which is the fundamental unit of mass in the theory of gravitation (it is the mass of hypothetical particles whose gravitational attraction for each other would be as strong as the electric force between two electrons separated by the same distance). But the Planck mass is about a hundred thousand trillion times larger than the Higgs mass. So, although the Higgs particle is so heavy that a giant particle collider was needed to create it, we still have to ask, why is the Higgs mass so small?

---

Jim Baggott suggested that I might add here some personal perspectives about the evolution of ideas in this field. I'll mention just two points.

As Baggott describes in Chapter 4, Philip Anderson argued early, before 1964, that massless Nambu-Goldstone particles were not a necessary consequence of symmetry breaking. So why were I and other particle theorists not convinced by Anderson's argument? It certainly did not reflect any judgment that Anderson did not have to be taken seriously. Of all the theorists who concerned themselves with condensed matter physics, no-one has seen more clearly than Anderson the importance of principles of symmetry, principles that have proved all-important in particle physics.

I think that Anderson's argument was generally discounted because it was based on analogies with phenomena like superconductivity which are non-relativistic (i.e., these are phenomena that occur in domains in which Einstein's special theory of relativity can be safely ignored). But the inevitability of massless Nambu-Goldstone particles had been shown, apparently rigorously, by Goldstone, Salam, and me, in a 1962 proof that relied on the manifest validity of relativity theory. Particle theorists were prepared to believe that Anderson was right in the non-relativistic context of superconductivity, but not in elementary particle theory, which necessarily incorporates relativity. The work of the 1964 papers made it clear that the proof by Goldstone, Salam, and myself did not apply to quantum theories with force-carrying particles, because although physical phenomena in such theories do satisfy the principle of relativity, the mathematical formulation of these theories in the context of quantum mechanics does not.

This problem with relativity was also the reason I was unable after 1967, despite strenuous efforts, to prove what Salam and I had conjectured, that nonsensical infinities that appeared in the electroweak theory all cancelled out, in the same way that similar infinities had already been shown to cancel in the quantum theory of electromagnetism alone. Relativity had been essential in demonstrating the cancellation of infinities in electromagnetism. The proof of cancellation by Gerard 't Hooft in 1971, described by Baggott in Chapter 5, used techniques that 't Hooft had worked out with Martinus Veltman, in which the principles of quantum mechanics are stretched to allow the theory to be formulated in a way that is consistent with relativity.

A second point: Baggott suggests in Chapter 4 that I did not include quarks in my 1967 paper proposing the electroweak theory because I was concerned about the problem that the theory might predict processes involving so-called 'strange' particles that were not in fact observed. I wish that my reason had been that specific. Rather, I did not include quarks in the theory simply because in 1967 I just did not believe in quarks. No-one had ever observed a quark, and it was hard to believe that this was because quarks are much heavier than observed particles like protons and neutrons, when these observed particles were supposed to be made of quarks.

Like many other theorists, I did not fully accept the existence of quarks until the 1973 work of David Gross and Frank Wilczek, and David Politzer. They showed that in the theory of quarks and strong nuclear forces known as quantum chromodynamics, the strong force gets weaker with decreasing distance. It then occurred to some of us that in that case the strong force between quarks would counter-intuitively get stronger as the quarks get farther apart, perhaps so much so as to prevent quarks from ever being separated from one another. There still is no proof of this, but it is generally believed. Quantum chromodynamics is by now a very well tested theory, and yet no-one has ever seen an isolated quark.

I was very glad to see this book begin in the early twentieth century with Emmy Noether, who realized before anyone else the importance of symmetry principles in nature. This helps to remind us that the work of scientists today is just the latest step in a grand tradition, of trying to guess how nature works, always subjecting our guesses to the test of experiment. Jim Baggott's book should give the reader some of the flavour of this historic enterprise.

Steven Weinberg
6 July 2012

# PROLOGUE
## *Form and Substance*

What is the world made of?

Simple questions such as this have been teasing the human intellect for as long as humankind has been capable of rational thought. For sure, the way we ask this question today has become much more elaborate and sophisticated, and the answers have become much more complex and costly to provide. But, make no mistake, at heart the question remains a very simple one.

Two and a half thousand years ago, all the Ancient Greek philosophers had to go on was their sense of beauty and harmony in nature and their powers of logical reasoning and imagination applied to the things they could perceive with their unaided senses. With hindsight, it is quite extraordinary just how much they were able to figure out.

The Greeks were careful to distinguish between form and substance. The world is made of material substance which can take a variety of different forms. The fifth century BC Sicilian philosopher Empedocles suggested that this variety could be reduced to four basic forms, what we know today as 'classical elements'. These were earth, air, fire, and water. The elements were judged to be eternal and indestructible, joined together in rather romantic combinations through the attractive force of Love and split apart through the repulsive force of Strife, to make up everything in the world.

Another school of thought originating with the fifth century BC philosopher Leucippus (and most closely associated with his pupil, Democritus) held that the world consists of tiny, indivisible, indestructible material particles (called atoms) and empty space (void). The atoms represented the building blocks of all material substance, responsible for all matter. Atoms were necessary as a matter of principle, so Leucippus argued, because substance surely could not be divided indefinitely. If this were possible, then we would be able to divide substance endlessly into nothing, in apparent contradiction with what seemed to be an unassailable law of the conservation of matter.

About a century later, Plato developed a theory which explained how atoms (the substance) are structured to make up the four elements (the forms). He represented each of the four elements by a geometrical (or 'Platonic') solid, and argued in the *Timaeus* that the faces of each solid could be further decomposed into systems of triangles, representing the elements' constituent atoms. Rearrange the patterns of triangles – rearrange the atoms – and it was possible to convert one element into another and combine elements to produce new forms.*

It seems logical that there should be some ultimate constituents, some undeniable reality that underpins the world we see around us and which lends it form and shape. If matter is endlessly divisible, then we would reach a point where the constituents themselves become rather ephemeral – to the

* See Plato, *Timaeus and Critias*, Penguin, London (1971), pp. 73–87. Plato built air, fire, and water from one type of triangle and earth, Penguin, London (1971), pp. 73–87. Plato built air, fire, and water from one type of triangle and earth from another. Consequently, Plato argued that it is not possible to transform earth into other elements.

point of non-existence. Then there would be no building blocks, and all we would be left with are interactions between indefinable, insubstantial phantoms which give rise to the *appearance* of substance.

Unpalatable it may be but, to a large extent, this is precisely what modern physics has shown to be true. Mass, we now believe, is not an inherent property or 'primary' quality of the ultimate building blocks of nature. In fact, there is no such thing as mass. Mass is constructed entirely from the energy of interactions involving naturally massless elementary particles.

The physicists kept dividing, and in the end found nothing at all.

---

It was not until the development of a formal experimental philosophy in the early seventeenth century that it became possible to transcend the kind of speculative thinking that had characterized the theories of the Ancient Greeks. The old philosophy had tried to intuit the nature of material substance from observations contaminated with prejudices about how the world *ought* to be. The new scientists now tinkered with nature itself, teasing out evidence about how the world *really is*.

The questions were still primarily concerned with the nature of form and substance. The concept of mass – a measure of the *amount* of matter as manifested in the dynamical movements of objects – became central to our understanding of substance. An object's resistance to acceleration is interpreted as inertial mass. When kicked with the same force, a small object will accelerate much faster than a large one.

An object's ability to generate a gravitational field is interpreted as gravitational mass. The force of gravity generated by the moon is weaker than the force generated by the earth, because the moon is smaller and so possesses a smaller gravitational mass. Inertial and gravitational mass are empirically identical, although there is no compelling theoretical reason why this should be so.

The scientists also exposed the secrets of nature's great variety of form. The fundamental Greek 'element' water was found to consist not of geometrical solids composed of triangles, as Plato had surmised, but of molecules composed of atoms of the chemical elements hydrogen and oxygen, in a combination we write today as $H_2O$.

This more modern use of the word 'atom' initially evoked the interpretation lent to it by the Greeks, as an indivisible building block of matter. But, even as the reality of atoms was being hotly debated, in 1897 English physicist Joseph John Thompson was discovering the negatively charged electron. It seemed that atoms, in their turn, possessed constituent, subatomic, parts.

Thompson's discovery was followed in the years 1909–11 by experiments in the Manchester laboratory of New Zealander Ernest Rutherford. These experiments showed that atoms consist, for the most part, of empty space. At the centre of the atom sits a tiny, positively charged nucleus, around which the negatively charged electrons orbit much like planets orbit the sun. Most of the mass of the atoms that make up the elements of material substance is concentrated in their atomic nuclei. It is therefore in the nucleus that form and substance come together.

This 'planetary' model of the atom remains a compelling visual metaphor even today. But it was immediately obvious to the physicists of the time that such a model actually makes no sense. Such planetary atoms were expected to be inherently unstable. Unlike planets moving around the sun, electrically charged particles moving in an electric field radiate energy in the form of electromagnetic waves. Such planetary electrons would exhaust their energy within a fraction of a second, and the internal architecture of the atom would then collapse.

The solution to this particular puzzle emerged in the guise of quantum mechanics in the early 1920s. The electron is not just a particle – which we might visualize as a tiny ball of negatively charged matter – it is simultaneously both wave *and* particle. It is not 'here' or 'there', as might be expected of a localized bit of stuff, but literally 'everywhere' within the confines of its ghostly, delocalized wavefunction. Electrons do not orbit the nucleus as such. Instead their wavefunctions form characteristic three-dimensional patterns – which we call 'orbitals' – in the space around the nucleus. The mathematical form of each orbital relates the *probability* of finding the now wholly mysterious electron at specific locations – 'here' or 'there' – inside the atom (see Figure 1).

The quantum revolution was a time of unprecedented fertility in both theoretical and experimental physics. When in 1927 English physicist Paul Dirac combined quantum mechanics with Albert Einstein's special theory of relativity, out popped a new property called *electron spin*. This was a property already known to experimentalists, and tentatively interpreted in terms of an electron spinning on its axis like a

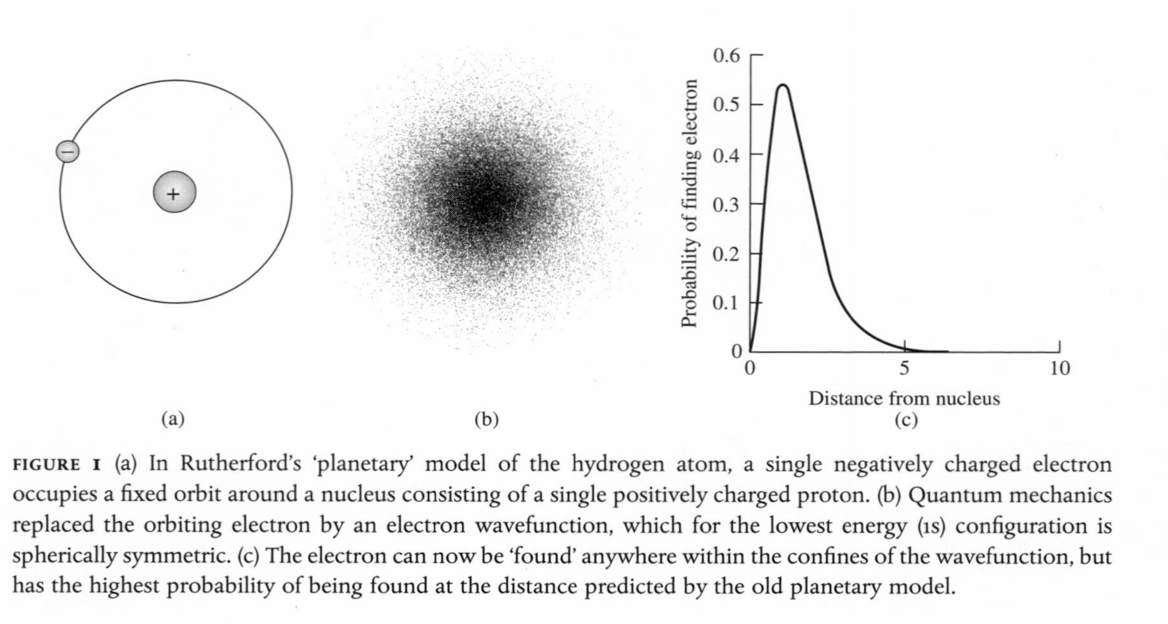


FIGURE 1 (a) In Rutherford's 'planetary' model of the hydrogen atom, a single negatively charged electron occupies a fixed orbit around a nucleus consisting of a single positively charged proton. (b) Quantum mechanics replaced the orbiting electron by an electron wavefunction, which for the lowest energy (1s) configuration is spherically symmetric. (c) The electron can now be 'found' anywhere within the confines of the wavefunction, but has the highest probability of being found at the distance predicted by the old planetary model.

spinning top, much as the earth rotates on its axis as it orbits the sun (see Figure 2).

But this was another visual metaphor that was quickly found to have no foundation in reality. Today, we interpret electron spin as a purely 'relativistic' quantum effect, in which electrons may take up one of two possible 'orientations', which we call spin-up and spin-down. These are not orientations along specific directions in conventional, three-dimensional space, but orientations in a 'spin-space' which has only two dimensions – up or down.

Each orbital in an atom was found to contain two – and only two – electrons. This is Austrian physicist Wolfgang Pauli's famous *exclusion principle*, which he developed in 1925 and which states that electrons are forbidden from occupying the same quantum state. The principle derives from the

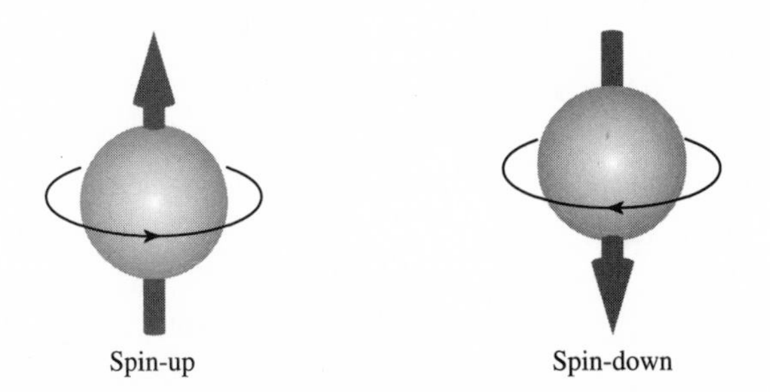


FIGURE 2 In 1927 Dirac combined quantum mechanics and Einstein's special theory of relativity to create a fully 'relativistic' quantum theory. Out popped the property of electron spin, imagined as though the negatively charged electron were literally spinning on its axis thereby generating a small, local magnetic field. Today we think of electron spin simply in terms of its possible orientations – spin-up and spin-down.

mathematical form of the wavefunction for any composite state consisting of two or more electrons. If the composite state were assumed to be created with two electrons which have precisely the same physical characteristics, then the wavefunction has zero amplitude – such a state could not exist. For the wavefunction to exist with a non-zero amplitude, then the two electrons must somehow be different. In an atomic orbital, this means that one electron must have a spin-up orientation and one must have a spin-down orientation. In other words, their spins must be *paired*.

It is wise to resist the temptation to imagine what these different orientations might actually look like. Their effects are real enough, however. Spin determines the amount of angular momentum carried by the electron – the momentum associated with the 'rotational' motion of its spin. Spin also governs how the electron interacts with a magnetic field, effects that can be studied in detail in the laboratory. But in quantum mechanics we appear to have crossed the threshold between what we can know of the origin of these effects, and what we cannot.

Dirac's relativistic quantum theory of the electron also yielded up twice as many solutions as he thought he had needed. Two of the solutions correspond to the spin-up and spin-down orientations of the electron. So what did the other two solutions correspond to? He had some ideas of his own, but finally conceded in 1931 that they had to represent the spin-up and spin-down orientations of a previously unknown positive electron. Dirac had discovered anti-matter. The 'positron', the anti-particle of the electron, was subsequently found in experiments on cosmic rays, formed high in the earth's atmosphere by collisions involving high-energy particles.

In 1932 it seemed that the final piece of the puzzle had been found. English physicist James Chadwick discovered the neutron, an electrically neutral particle which sits snugly alongside the positively charged proton inside the atomic nucleus. It seemed that physicists now had all the ingredients to formulate a definitive answer to our opening question.

The answer went something like this. All the material substance in the world is made of chemical elements. These elements come in a great variety of forms which make up the periodic table, from the lightest, hydrogen, to the heaviest-known, naturally occurring element, uranium.*

Each element consists of atoms. Each atom consists of a nucleus composed of varying numbers of positively charged protons and electrically neutral neutrons. Each element is characterized by the number of protons in the nuclei of its atoms. Hydrogen has one, helium two, lithium three, and so on, to uranium, which has 92.

Surrounding the nucleus are negatively charged electrons, in numbers which balance the numbers of protons, so that overall the atom is electrically neutral. Each electron can take either a spin-up or spin-down orientation and each orbital can accommodate two electrons provided their spins are paired.

It is a very comprehensive answer. With fundamental building blocks of protons, neutrons, and electrons and Pauli's

* There are elements heavier than uranium, but these do not occur in nature. They are inherently unstable and must therefore be produced artificially in a laboratory or a nuclear reactor. Plutonium is perhaps the best-known example.

exclusion principle, we can explain why the periodic table has the structure that it has. We can explain why matter has shape and density. We can explain the existence of isotopes – atoms with the same numbers of protons but different numbers of neutrons in their nuclei. With a little effort, we can explain all of chemistry, biochemistry, and materials science.

In this description, mass is no real mystery. The mass of all material substance can be traced back to its constituent protons and neutrons, which account for about 99 per cent of the mass of every atom.

Imagine a small cube of ice, formed from triply distilled water. Its sides measure 2.7 centimetres in length, or a little over an inch. Pick it up. It's cold and slippery. It's not heavy, but you are conscious of its weight in the palm of your hand. So, where does the mass of the ice cube reside?

The *molecular weight* of water is simply calculated from the total number of protons and neutrons in the nuclei of the two atoms of hydrogen and one atom of oxygen that make up $H_2O$. The nucleus of each hydrogen atom consists of just one proton, and the nucleus of the oxygen atom contains eight protons and eight neutrons, making 18 'nucleons' in total. The cube of pure ice you hold in your hand will weigh about 18 grams,* equal to the molecular weight in grams. The cube therefore represents a standard measure of solid water known as a 'mole'.

* The density of pure ice at 0°C is 0.9167 grams per cubic centimetre. The ice cube has a volume of about 19.7 cubic centimetres, so its mass is a little over 18 grams.

We know that a mole of substance contains a fixed number of the atoms or molecules that make up that substance. This is Avogadro's number, a little over six hundred billion trillion ($6 \times 10^{23}$). Here, then, is the answer. The weight of the ice cube that you feel in the palm of your hand is the combined result of the masses of six hundred billion trillion molecules of $H_2O$, or about 10,800 billion trillion protons and neutrons (see Figure 3).*

It had to be accepted that atoms were no longer indestructible, as the Greeks had once thought. Atoms could be transmuted, turned from one form into another. In 1905, Einstein

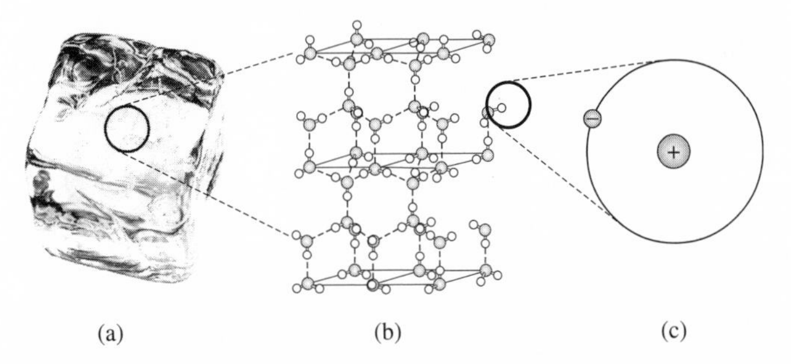


**FIGURE 3** A cube of ice measuring 2.7 centimetres in length will weigh about 18 grams (a). It consists of a lattice structure containing a little over six hundred billion trillion molecules of water, $H_2O$ (b). Each atom of oxygen contains eight protons and eight neutrons and each hydrogen atom contains one proton (c). The cube of ice therefore contains about 10,800 billion trillion protons and neutrons.

* Of course, we need to be careful to distinguish between weight and mass. The ice cube *weighs* 18 grams on earth but it weighs a lot less on the moon and nothing at all in orbit around the earth. Its *mass*, however, remains firmly fixed. By convention, we set the mass to be equal to its earthly weight.

had used his special theory of relativity to show that mass and energy are equivalent, through what was to become the world's most famous scientific equation, $E = mc^2$: energy is equal to mass multiplied by the speed of light squared. However, far from undermining the concept of mass, the notion that mass represents a vast reservoir of energy somehow made it even more substantial.

Substantial, but not immutable. Einstein showed that matter (mass) is not conserved – it can be converted to energy. When an atom of uranium-235 is fissioned by bombardment with a fast neutron, about one-fifth of the mass of a single proton is converted into energy in the resulting nuclear reaction. When scaled up to a 56-kilogram bomb core of 90 per cent pure uranium-235, the amount of mass-energy released was sufficient to destroy utterly the Japanese city of Hiroshima in August 1945.

But Einstein was actually chasing a deeper truth. There was a clue in the title of his 1905 paper: 'Does the inertia of a body depend on its energy content?'[1] Einstein had understood that $E = mc^2$ really means that $m = E/c^2$: all inertial mass is just another form of energy.* The profound implications of this observation would not become apparent for another sixty years.

---

By the mid-1930s, fundamental building blocks of protons, neutrons, and electrons seemed to provide a comprehensive answer to our opening question. But there was a problem. It

* In fact, the equation $E = mc^2$ does not appear in this form in Einstein's paper.

had been known since the late nineteenth century that isotopes of certain elements are unstable. They are radioactive: their nuclei disintegrate spontaneously in a series of nuclear reactions.

There are different kinds of radioactivity. One kind, which was called beta-radioactivity by Rutherford in 1899, involves the transformation of a neutron in a nucleus into a proton, accompanied by the ejection of a high-speed electron (a 'beta-particle'). This is a natural form of alchemy: changing the number of protons in the nucleus necessarily changes its chemical identity.*

Beta-radioactivity implied that the neutron is an unstable, composite particle, and so not really 'fundamental' at all. There was also a problem with the balance of energy in this process. The theoretical energy released by the transformation of a proton inside the nucleus could not all be accounted for by the energy of the emitted electron. In 1930 Pauli had felt that he had no choice but to propose that the energy 'missing' in the reaction was being carried away by an as yet unobserved, light, electrically neutral particle which eventually came to be called a *neutrino* ('small neutral one'). At the time it was judged that it would be impossible to detect such a particle, but it was first discovered in 1956.

It was time to take stock. This much was clear. Matter relies on *force* to hold it together. Aside from the force of gravity, which acts universally on all matter, there were now judged to be three other kinds of force at play within the atom itself.

* Fortunately for the value of the world's gold reserves, this does not provide a cheap way to transform base metals into gold.

The interactions between electrically charged particles are derived from the force of electromagnetism, well known from the pioneering work of nineteenth century physicists which, among many notable achievements, laid the foundations for the power industry. A fully relativistic quantum theory of the electromagnetic field, called quantum electrodynamics (QED), was worked out in 1948 by American physicists Richard Feynman and Julian Schwinger, and Japanese physicist Sin-Itiro Tomonaga. In QED, the forces of attraction and repulsion between electrically charged particles are 'carried' by so-called force particles.

For example, as two electrons approach each other, they exchange a force particle which causes them to be repelled

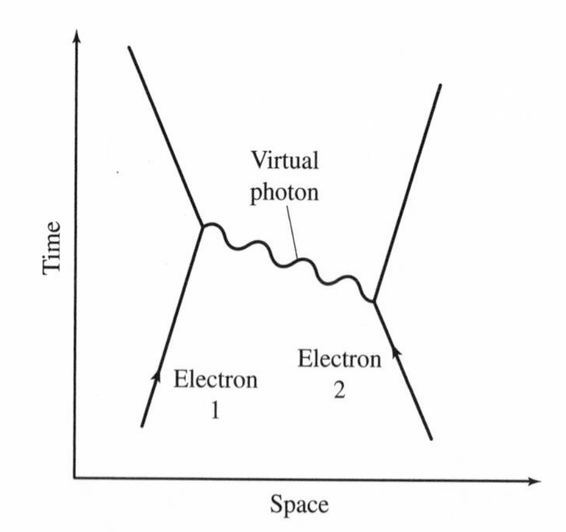


**FIGURE 4** Representation of the interaction between two electrons as described by quantum electrodynamics. The electromagnetic force of repulsion between the two negatively charged electrons involves the exchange of a virtual photon at the point of closest approach. The photon is 'virtual' as it is not visible during the interaction.

(see Figure 4). The force carrier of the electromagnetic field is the photon, the quantum particle which makes up ordinary light. QED was quickly established as a theory of unprecedented predictive power.

There were yet another two forces to contend with. Electromagnetism could not explain how protons and neutrons bind together inside atomic nuclei, nor could it explain the interactions associated with beta-radioactive decay. These interactions work at such different energy scales that no single force can accommodate both. It was recognized that two forces were needed, a 'strong' nuclear force responsible for holding atomic nuclei together and a 'weak' nuclear force governing certain nuclear transformations.

---

This brings us to the period in the history of physics to be described in this book. A further 60 years of theoretical and experimental particle physics have brought us to the Standard Model, a collection of fundamental quantum field theories which describe all matter and all the forces between material particles with the exception of gravity. The easiest way to appreciate what the Standard Model is and what it means for our understanding of the material world is to take a quick tour through its history.

Our journey begins in 1915, in the quiet German university town of Göttingen.

# PART I
# Invention

# 1

# The Poetry of Logical Ideas

*In which German mathematician Emmy Noether discovers the relationship between conservation laws and the deep symmetries of nature*

Perhaps we might agree that one of the aims of science is to explain what the world is made of and why it is the way it is. It seeks to do this by elucidating the fundamental constituents of matter and the laws of nature that govern its behaviour.

If we can agree on this point, we would then have to admit that not all 'laws' are the same. Not all laws are truly fundamental. In the seventeenth century, Johannes Kepler toiled for many years with Tycho Brahe's painstakingly wrought astronomical data and eventually devised three laws that govern the motions of the planets around the sun. These laws are very powerful, but they belie a more fundamental explanation, a reason why the planets orbit the sun in the way they do. Isaac Newton's law of universal gravitation provided just such an explanation. Newton's law stood firm for another two hundred years before it was eventually replaced by the interplay of matter and curved space-time in Einstein's general theory of relativity.

So, what are the 'fundamental' laws? This is perhaps not so difficult to answer. Much of what we understand about the nature of our world is founded on some deceptively simple laws of *conservation*. The Ancient Greeks believed that matter is conserved. They were almost right. Einstein later taught that matter can be reduced to energy, and from energy can spring matter.

Matter (in the form of material substance) is not conserved, but mass-energy is. No matter how hard we try, we cannot make or destroy energy. We can only convert it from one kind to another. In every physical interaction of every conceivable description, energy is conserved.

So too is linear momentum, the mass of an object multiplied by its velocity in a straight line. At first, this does not seem to be consistent with common experience. A popular theme-park ride shoots thrill-seekers at high velocity horizontally along a track.* The track loops-the-loop. The car carrying the passengers then climbs a steep ramp, losing its momentum before slowing to a halt. Gravity pulls it back down the ramp. The car picks up momentum and loops-the-loop *backwards* before finally coming to rest. Now, it seems quite clear that linear momentum is not conserved as the car climbs the ramp and comes to a stop.

But there is a bigger picture here. As the car loses momentum, the rest of the world beneath it and to which it is

* I enjoyed just such a ride whilst working as a postdoctoral researcher in California in the early 1980s. I think it was called the 'Tidal Wave'.

attached imperceptibly gains momentum, such that momentum is conserved.

So too is angular momentum, the momentum of rotating bodies, calculated as the linear momentum multiplied by the distance from the centre of the rotation. A figure-skater enters a spin with arms and one leg outstretched. As she draws her arms and leg back towards her centre of mass, she reduces the distance from her centre of rotation, and she spins faster. This is the conservation of angular momentum in action.

As the example of linear momentum shows, these laws are hardly intuitive. They were hinted at for many centuries but, to articulate a law of conservation, it is first necessary to be clear about the quantity that is being conserved. And the concept of energy was not properly formalized and understood until the nineteenth century.

The conservation laws as they are presented today represent the culmination of centuries of hit-and-miss experimentation and theorizing. Though fundamental, there is a sense in which these laws are nevertheless empirical – they derive from observations and experiments rather than from some deep, underlying theoretical model of the world. Could there be some deeper principle from which the conservation of energy and momentum would automatically result?

In 1915, German mathematician Amalie Emmy Noether certainly thought so.

---

Noether was born in Erlangen, in Bavaria, in March 1882. Her father, Max, was a mathematician at the University of Erlangen and in 1900 Emmy became one of only two female students to attend the university. Like all academic institutions in Germany at that time, the university did not wish to encourage female students and Emmy was obliged first to seek permission from her lecturers before attending their classes.

After graduating in the summer of 1903, she spent the winter months at the University of Göttingen. Here she was exposed to lectures delivered by some of Germany's leading mathematicians, including David Hilbert and Felix Klein. She then returned to Erlangen to work on her dissertation, and in 1908 she became an unpaid lecturer at the university.

She developed an interest in Hilbert's work and published several papers extending some of his methods in abstract algebra. Both Hilbert and Klein were impressed, and in early 1915 they sought to bring her back to Göttingen to join the faculty.

They met with stubborn resistance.

'What will our soldiers think when they return to the university and find they are required to learn at the feet of a woman?' argued the conservatives on the faculty.

'I do not see that the sex of the candidate is an argument against her admission as a Privatdozent [an assistant professor],' countered Hilbert. 'After all, we are a university, not a bath house.'[1]

Hilbert prevailed, and Noether moved to Göttingen in April 1915.

It was shortly after arriving in Göttingen that Noether formulated what was to become one of the most famous theorems in physics.

---

Noether deduced that the principles of the conservation of physical quantities such as energy and momentum can be traced to the behaviour of the laws describing them in relation to the operation of certain continuous symmetry transformations. The conservation laws are manifestations of the deep symmetries of nature.

We tend to think of symmetry in terms of mirror reflections: left–right, top–bottom, front–back. We say something is symmetrical if it looks the same on either side of some centre or axis of symmetry. In this case, a symmetry 'transformation' is the act of reflecting an object as though in a mirror. If the object is unchanged (or 'invariant') following such an act we say it is symmetrical.

To take one example, it seems that facial symmetry is deeply woven into our human perception of beauty and attractiveness, serving as a subliminal indicator of genetic quality. Those exalted as beautiful people tend to have more symmetrical faces and, generally speaking, we tend to want to mate with those we regard as beautiful (see Figure 5).*

These examples of symmetry transformations are said to be 'discrete'. They require an instantaneous 'flipping' from one

* There is evidence to suggest that women's bodies actually become more symmetrical in the 24 hours prior to ovulation. See Brian Bates and John Cleese, *The Human Face*, BBC Books, London (2001), p. 149.

FIGURE 5 We tend to think of symmetry in terms of mirror reflections, and say that something is symmetrical if it looks the same on either side of some centre or axis of symmetry. Elizabeth Hurley demonstrates the relationship between facial symmetry and classical beauty.
*Source*: ©Peter Steffen/dpa/Corbis

perspective to another, such as left-to-right. The kinds of symmetry transformations involved in Noether's theorem are very different. They involve continuous, gradual changes, such as a continuous rotation in a circle. It seems blindingly obvious that if we rotate a circle through an infinitesimally small angle measured from its centre, then the circle appears unchanged. The circle is symmetric to continuous rotational transformations. A square is not symmetric in this same sense. It is rather symmetric to discrete rotations through 90° (Figure 6).

Noether's theorem connects each conservation law with a continuous symmetry transformation. She found that the

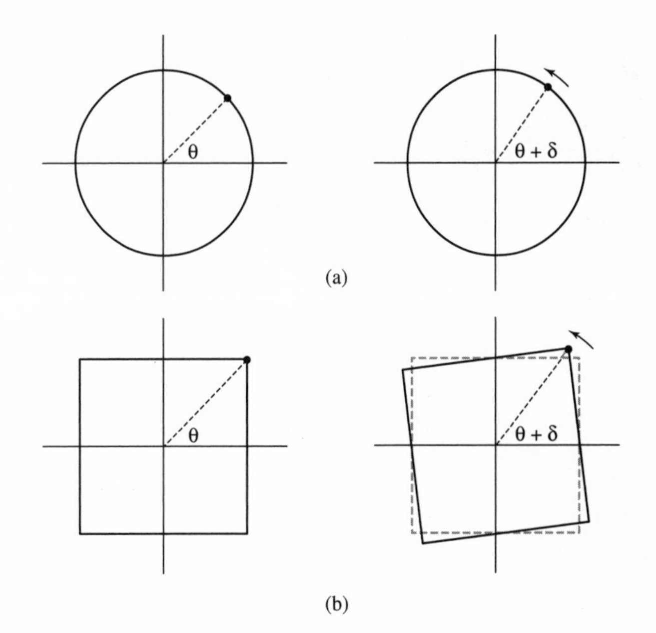


**FIGURE 6** Continuous symmetry transformations involve small, incremental changes to a continuous variable, such as a distance or an angle. (a) When we rotate a circle through a small angle (δ), the circle appears unchanged (or 'invariant') and we say that it is symmetric to such transformations. (b) In contrast, a square is not symmetric in this same sense. A square is, instead, symmetric to *discrete* rotations through 90°.

laws governing energy are invariant to continuous changes or 'translations' in *time*. In other words, the mathematical relationships which describe the dynamics of energy in a physical system at some time $t$ are exactly the same an infinitesimally short time later.

This means that these laws do not change with time, which is precisely what we might expect from relationships between physical quantities that we wish to elevate to the status of fundamental 'laws'. These laws are the same yesterday, today, and tomorrow, which is immensely reassuring. If the laws describing energy do not change with time, then energy *must* be conserved.

For linear momentum, Noether found the laws to be invariant to continuous *translations in space*. The laws governing conservation of linear momentum do not depend on any specific location in space. They are the same here, there, and everywhere. For angular momentum, the laws are invariant to rotational symmetry transformations, as in the example of a circle described above. They are the same irrespective of the *angle of direction* measured from the centre of the rotation.

The logic by which Noether arrived at her theorem goes something like this. In physics, there are certain quantities that appear from careful observation and experiment to be conserved. After much effort, physicists deduced the laws governing these quantities. These laws are found to be invariant to certain continuous symmetry transformations. Such invariance means that the quantities so governed are *required* to be conserved.

This logic could now be turned on its head. Suppose there is a physical quantity which appears to be conserved but for which the laws governing its behaviour have yet to be properly elucidated. If the physical quantity is indeed conserved, then the laws – whatever they are – must be invariant to a particular continuous symmetry transformation. If we can discover what this symmetry is, then we are well on the way to identifying the laws.

Inverting Noether's logic offers a way of avoiding considerable hit-and-miss theorizing. Physicists were presented with an approach to identifying laws that helped to rule out whole varieties of possible mathematical structures. Finding the symmetry underlying the physical quantity offers a shortcut to the answer.

---

There was indeed a physical quantity that appeared to be rigorously conserved but for which the appropriate laws had yet to be deduced. This was electric charge.

The phenomenon of static electricity was known to the philosophers of Ancient Greece. They found that they could generate electric charges and even sparks by rubbing substances such as amber with fur. The scientific study of electricity has a long and illustrious history, with many participants. But it was English physicist Michael Faraday, working at the Royal Institution in London, who synthesized the multitude of observations and experiments into a single, coherent understanding of the nature of electric charge. The result of much experimentation led to the inescapable conclusion that electric change cannot be created or destroyed in any physical or chemical change. Charge is conserved.

There was no shortage of laws or rules governing electric charge and its rather mysterious connection with magnetism – Coulomb's law, Gauss's law, Ampère's law, Biot-Savart's rule, Faraday's law, and so on. In the early 1860s, Scottish physicist James Clerk Maxwell did for electromagnetism what Newton had done for the theory of planetary motion. He provided a bold theoretical synthesis to parallel Faraday's experimental unification. Maxwell's beautiful equations tied in an intimate embrace the electric and magnetic fields generated by a moving electric charge.*

The equations also demonstrated that all electromagnetic radiation – including light – can be described as a wave motion with a speed that can be calculated from known physical constants. These are the permittivity of free space, which is a measure of the ability of empty space to transmit or 'permit' an electric field generated by an electric charge, and the permeability of free space, a measure of the ability of empty space to develop a magnetic field surrounding a moving electric charge. When Maxwell combined these constants in the way dictated by his new electromagnetic theory, the result he obtained for the speed of his 'waves of electromagnetism' was precisely the speed of light.

But Maxwell's equations deal with the *fields* generated by electric charge, not the charge itself. These are intimately related, but the equations give no basis in principle for understanding the origin of charge conservation. In the light of Noether's theorem, the search for the laws governing electric charge became a search for the underlying continuous symmetry transformation to which the laws are invariant.

* Time to explain what we mean here by 'fields'. The field associated with a force such as gravity or electromagnetism has both a magnitude and a direction at every point in the space surrounding the object that generates it. You can detect this field by placing in it another object that is susceptible to the force. Pick up any object (preferably nothing breakable) and drop it. The object's response is governed by the magnitude and direction of the gravitational field at the precise point where you let go. The object feels the force, and falls to the ground.

The search was picked up by German mathematician Hermann Weyl.

Born in 1885 in Elmshorn, a small town near Hamburg, Weyl had secured his doctorate under Hilbert's supervision at Göttingen in 1908. He had then taken up a professorship at the Eidgenossische Technische Hochschule (ETH) in Zurich, where he met Albert Einstein and became fascinated by problems in mathematical physics.

In developing his general theory of relativity in 1915, Einstein had eliminated any sense of absolute space and time. Instead, he argued, physics should depend only on the distances between points and the curvature of space-time at each point. This is Einstein's principle of *general covariance,* and the theory of gravitation that results is invariant to arbitrary changes of coordinate system. In other words, although there are natural physical laws, there is no 'natural' coordinate system of the universe. We invent coordinate systems to help describe the physics but the laws themselves should not (and do not) depend on these arbitrary choices.

There are two ways we can change the coordinate system. We can make a *global* change, applied uniformly at all points in space and time. An example of such a global symmetry transformation is a uniform shift in the lines of latitude and longitude used by cartographers to map the surface of the earth. So long as the change is uniform and applied consistently across the globe, this makes no difference to our ability to navigate from one place to another.

But changes can also be made *locally,* with different changes to the coordinates at different points in space-time. For example, in one particular part of space we could choose to rotate the axes of our coordinate system through a small angle, at the same time changing the scale. Provided this change is translated through to the measures of differences in position and differences in time, this makes no difference to the predictions of general relativity. General covariance is therefore an example of invariance to local symmetry transformations.

Weyl thought long and hard about Noether's theorem and worked on the theory of groups of continuous symmetry transformations called Lie groups, named for the nineteenth century Norwegian mathematician Sophus Lie. In 1918 he concluded that the conservation laws are related to local symmetry transformations to which he gave the generic name *gauge symmetry*, an unfortunately rather obscure term. Guided by Einstein's work, he was thinking of symmetry in relation to distances between points in space-time, as in the example of a train running on tracks with a fixed gauge.

He found that by generalizing the principle of general covariance to one of gauge invariance, he could use Einstein's theory as a basis for the derivation of Maxwell's equations for electromagnetism. What he had discovered appeared to be a theory that could unify the two forces then known to science – electromagnetism and gravity. The invariance identified with the conservation laws would then be related to arbitrary changes in the 'gauge' of the fields involved. In this way, Weyl hoped to demonstrate the conservation of energy, linear and angular momentum, *and* electric charge.

Weyl initially ascribed his gauge invariance to space itself. But, as Einstein quickly pointed out, this meant that the measured lengths of rods and the readings of clocks would come to depend on their recent history. A clock moved around a room would no longer keep time correctly. Einstein wrote to Weyl, complaining: 'Apart from the agreement with reality, [your theory] is at any rate a grandiose achievement of the mind.'[2]

Weyl was disturbed by this criticism, but accepted that Einstein's intuition in these matters was normally reliable. Weyl abandoned his theory.

---

Austrian physicist Erwin Schrödinger joined the faculty at the University of Zurich three years later, in 1921. He was diagnosed with suspected pulmonary tuberculosis just a few months later. He was ordered to take a complete rest cure. He and his wife Anny retreated to a villa in the Alpine resort of Arosa, near the fashionable ski resort of Davos, where they stayed for nine months.

As Anny nursed him back to health, he pondered on the significance of Weyl's gauge symmetry and, specifically, a periodic 'gauge factor' which appeared in Weyl's theory. In 1913, Danish physicist Niels Bohr had published details of a theory of atomic structure in which electrons are required to orbit the nucleus at fixed energies characterized by their 'quantum numbers'. These integral numbers govern the energies of the orbits, increasing in linear sequence (1, 2, 3,...) from the innermost to the outermost orbit. At the time, their origin was a complete mystery.

Schrödinger was struck by the possibility that there might be a connection between the periodicity implied by Weyl's gauge factor and the periodicity implied by Bohr's quantized atomic orbits. He examined a couple of possible forms for the gauge factor, including one containing a complex number, formed by multiplying a real number by the 'imaginary' number *i*, the square root of –1.* In a paper published in 1922 he suggested that this connection had a deep physical significance. But this was a vague intuition. The real significance of the connection would elude him until he studied the 1924 doctoral thesis of French physicist Louis de Broglie.

De Broglie had suggested that, just as electromagnetic waves can appear to behave like particles,† so perhaps particles like electrons could sometimes behave like waves. Whatever they were, these 'matter waves' could not be considered to be in any way like more familiar wave phenomena, such as sound waves or water waves. De Broglie concluded that the matter wave: 'represents a spatial distribution of *phase*, that is to say, it is a "*phase wave*".'[3]‡

Schrödinger was set to thinking: what would the electron look like if it was described mathematically as a wave? At

* This is 'imaginary' only in the sense that it is not possible to calculate the square root of –1. When squared, any positive or negative number will always give a positive answer. But even though the square root of –1 doesn't exist, this doesn't stop mathematicians from using it. Thus, the square root of any negative number can be expressed in terms of *i*. For example, the square root of –25 is 5*i*, which is called a complex or imaginary number.

† These were called 'light-quanta' by Einstein in 1905. Today we call them photons.

‡ A familiar example of a phase wave is provided by a 'Mexican' wave travelling around a sports stadium. The wave is created by the motions of individual spectators as they change positions, from standing with their arms raised (the phase 'peak') to sitting in their seats (the phase 'trough'). The phase wave is the result of the coordinated movements of the spectators, and can travel around the stadium a lot faster than can the individual spectators who support it.

Christmas 1925 he retreated once more to Arosa. His relationship with his wife was at an all-time low, so he chose to invite an old girlfriend from Vienna to join him. He also took with him his notes on de Broglie's thesis. When he returned on 8 January 1926, he had discovered *wave mechanics*, a theory which describes the electron as a wave and the orbits of Bohr's atomic theory in terms of electron 'wavefunctions'.

It was now possible to make the connection. One example of a Lie group is the symmetry group U(1), referred to as the unitary group of transformations of one complex variable. This involves symmetry transformations that are, in many ways, entirely analogous to those involving continuous rotation in a circle. But whereas a circle is drawn in a two-dimensional plane formed from real dimensions, the transformations of the symmetry group U(1) involve rotations in a two-dimensional *complex plane*. This is formed from two 'real' dimensions, with one of them multiplied by *i*.

Another way of representing this symmetry group is in terms of continuous transformations of the *phase angle* of a sinusoidal wave (see Figure 7). Different phase angles correspond to different amplitudes of the wave in its peak–trough cycle. Weyl's gauge symmetry is preserved if changes in the phase of the electron wavefunction are matched by changes in its accompanying electromagnetic field. The conservation of electric charge can be traced to the local phase symmetry of the electron wavefunction.

The connection between wave mechanics and Weyl's gauge theory was made explicit in 1927 by young German theorist Fritz London and Soviet physicist Vladimir Fock. Weyl recast

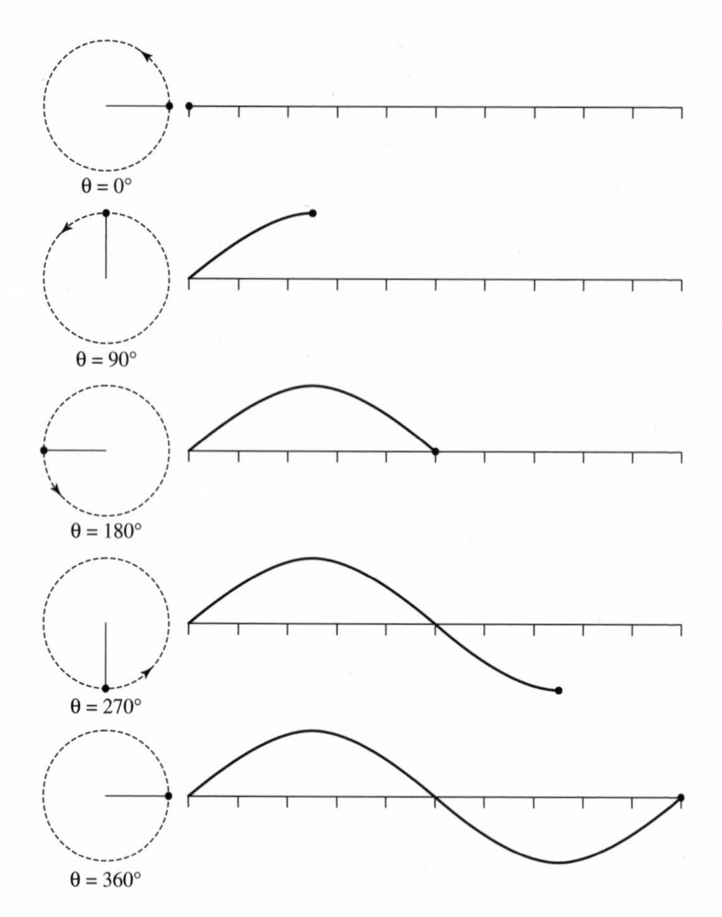


**FIGURE 7** The symmetry group U(1) is the unitary group of transformations of one complex variable. In a complex plane formed by one real axis and one imaginary axis, we can pinpoint any complex number on the circumference of the circle formed by rotating the line drawn from the origin to the point through the continuous angle, $\theta$, that this line makes with the real axis. There is a deep connection between this continuous symmetry and simple wave motion, in which the angle $\theta$ is a phase angle.

and extended his theory in the context of quantum mechanics in 1929.

---

De Broglie's wave–particle 'duality' implied that the electron was to be regarded as both wave and particle. But how could this be? Particles are localized bits of stuff, waves are delocalized disturbances in a medium (think of the ripples in a pond caused by the throw of a stone). Particles are 'here', waves are 'there and everywhere'.

One of the physical consequences of wave–particle duality is that we cannot measure the simultaneous position and momentum (specifically the speed and direction) of a quantum particle precisely. Think about it. If we can measure the precise position of a wave-particle this must mean that it is localized in space and time. It is 'here'. For a wave this is only possible if it is formed by combining a large number of wave forms of different frequencies, such that they add up to produce a wave which is large in one location in space and small everywhere else. This gives us the position, but at the cost of complete uncertainty in the wave frequency, because the wave must be composed of many waves with lots of different frequencies.

But in de Broglie's hypothesis, the inverse frequency of the wave is directly related to the particle momentum.* Uncertainty in frequency therefore means uncertainty in momentum.

* The de Broglie relationship is written $\lambda = h/p$, where $\lambda$ is the wavelength (related to the reciprocal of the frequency), $h$ is Planck's constant, and $p$ is the momentum. This means that $p = hc/v$, where c is the speed of light and $v$ is the frequency.

The converse is also true. If we want to be precise about the frequency of the wave, and hence the momentum of the particle, then we have to stick with a single wave with a single frequency. But then we can't localize it. The wave-particle remains spread out in space and we can no longer measure a precise position.

This uncertainty in position and momentum is the basis for German physicist Werner Heisenberg's famous uncertainty principle, discovered in 1927. It is a direct consequence of the duality of wave and particle behaviour in elementary quantum objects.

---

Weyl returned to Göttingen in 1930, taking the professorship vacated by the retiring Hilbert. He joined Noether, who had remained in Göttingen but for a short period of study leave at Moscow State University during the winter of 1928–29.

In January 1933 Adolf Hitler became Chancellor of Germany. A few months later Hitler's National Socialist government introduced the Law for the Reestablishment of the Career Civil Service, the first of four hundred such decrees. It provided a legal basis on which the Nazis could forbid Jews from holding positions in the Civil Service, including academic positions in German universities.

Weyl's wife was Jewish, and he left Germany to join Einstein at the Institute for Advanced Study in Princeton, New Jersey. Noether was Jewish, and she lost her position at Göttingen. She had never been promoted to the status of full professor.

She left for Bryn Mawr College, a liberal arts college in Pennsylvania. She died two years later, aged 53.

In an obituary that appeared in the *New York Times* shortly after her death, Einstein wrote:[4]

> In the judgment of the most competent living mathematicians, Fräulein Noether was the most significant creative mathematical genius thus far produced since the higher education of women began. In the realm of algebra, in which the most gifted mathematicians have been busy for centuries, she discovered methods which have proved of enormous importance in the development of the present-day younger generation of mathematicians. Pure mathematics is, in its way, the poetry of logical ideas. One seeks the most general ideas of operation which will bring together in a simple, logical, and unified form the largest possible circle of formal relationships. In this effort toward logical beauty spiritual formulas are discovered necessary for the deeper penetration into the laws of nature.

# 2

# Not a Sufficient Excuse

*In which Chen Ning Yang and Robert Mills try to develop a quantum field theory of the strong nuclear force and annoy Wolfgang Pauli*

When Dirac successfully combined quantum theory and Einstein's special theory of relativity in 1927, the result was electron spin and anti-matter. Dirac's equation was rightly regarded as an absolute wonder, but it was also quickly realized that this could not be the end of the story.

Physicists began to acknowledge that they needed a fully fledged relativistic theory of quantum electrodynamics, or QED. This would, in essence, be a quantum version of Maxwell's equations that conformed to Einstein's special theory of relativity. Such a theory would necessarily incorporate a quantum version of the electromagnetic field.

Some physicists believed that fields were more fundamental than particles. It was thought that a proper quantum field description should yield particles as the 'quanta' of the fields themselves, carrying the force from one interacting particle to another. It seemed clear that the photon was the field particle of the quantum electromagnetic field, created and destroyed when charged particles interact.

German physicist Werner Heisenberg and Austrian Wolfgang Pauli developed a version of just such a quantum field theory in 1929. But there was a big problem. The physicists found that they could not solve the field equations exactly. In other words, it was not possible to write down a solution to the field equations that took the form of a single, self-contained mathematical expression, applicable in all circumstances.

Heisenberg and Pauli had to resort to an alternative approach to solving the field equations based on a so-called perturbation expansion. In this approach, the equation is recast as the sum of a potentially infinite series of terms $- x^0 + x^1 + x^2 + x^3 + \cdots$ The series starts with a 'zeroth-order' (or zero-interaction) expression which can be solved exactly. To this is added additional (or perturbation) terms representing corrections to first-order ($x^1$), second-order ($x^2$), third-order ($x^3$), etc. In principle, each term in the expansion provides a smaller and smaller correction to the zeroth-order result, gradually bringing the calculation closer and closer to the actual result. The accuracy of the final result then depends simply on the number of perturbation terms included in the calculation.

But instead of finding smaller and smaller corrections, they found that some terms in the perturbation expansion mushroomed to infinity. When applied to the quantum field theory of the electron, these terms were identified to result from the electron's 'self-energy', a consequence of the electron interacting with its own electromagnetic field.

There was no obvious solution.

---

There the matter rested. James Chadwick discovered the neutron in 1932. In the years following this discovery Italian physicist Enrico Fermi used high-energy neutrons to bombard atoms of different chemical elements in search of interesting new physics. Puzzled by some of Fermi's results, German chemists Otto Hahn and Fritz Strassman studied the products from the neutron bombardment of uranium atoms. On Christmas Eve 1938, their even more puzzling results were discussed by Hahn's long-time collaborator Lise Meitner and her physicist nephew Otto Frisch, by now both exiled from Nazi Germany. Their animated discussion led to the discovery of nuclear fission.

It was a portentous discovery, reported in January 1939, just nine months before the beginning of the Second World War. Transformed from 'other-worldly eggheads' into the most important military resources of nation-states, the physicists now worked to turn the discovery of nuclear fission into the world's most dreadful weapon of war.

When the time finally came in 1947 to turn their attentions back to the problems that beset quantum electrodynamics, it was declared that theoretical physics had been in the doldrums for nearly two decades.

---

But there quickly followed another great burst of creativity. In June 1947 a group of leading American physicists gathered for a small, invitation-only conference at the Ram's Head Inn, a small clapboard hotel and inn on Shelter Island, at the eastern end of New York's Long Island.

It was an illustrious group. Among them were J. Robert Oppenheimer, the 'father' of the atom bomb, Hans Bethe, who had led the Theoretical Division at Los Alamos, Victor Weisskopf, Isidor Rabi, Edward Teller, John Van Vleck, John von Neumann, Willis Lamb, and Hendrik Kramers. A new generation of physicists was represented by John Wheeler, Abraham Pais, Richard Feynman, Julian Schwinger, and former Oppenheimer students Robert Serber and David Bohm. Einstein had been invited to attend but declined for reasons of ill-health.

The physicists heard of some disturbing new experimental results. One of the quantum states of the hydrogen atom was found to be shifted in energy slightly in relation to another, a phenomenon that came to be called the *Lamb shift*, after its discoverer, Willis Lamb. Dirac's theory predicted that both states should have precisely the same energy.

There was more. Rabi announced that a new measurement of the *g*-factor of the electron – a physical constant which reflects the strength of the interaction of an electron with a magnetic field – has a value of the order of 2.00244. Dirac's theory predicted a *g*-factor of exactly 2.

These were results that simply could not be predicted without a fully fledged QED. It seemed that although the theory was beset with problems inherent in its mathematical structure, nature itself had no problems with infinities. The physicists had to find a way around this somehow.

The discussion continued long into the night. The physicists split into groups of two and three, the corridors echoing their arguments, as they regained their passion for physics. Schwinger later remarked: 'It was the first time people who had all this physics pent up in them for five years could talk to each other without somebody peering over their shoulders and saying "Is this cleared?"'[1]

Then there came a glimmer of hope. Dutch physicist Kramers outlined a new approach to thinking about the mass of an electron in an electromagnetic field. He proposed to treat the self-energy of the electron as an additional contribution to its mass.

After the conference, Bethe returned to New York and took a train to Schenectady, where he was working as a part-time consultant to General Electric. As he sat on the train he played around with the equations of QED. The existing theories of QED predicted an infinite Lamb shift, a consequence of the electron's self-interaction. Bethe now followed Kramers' suggestion and identified the infinite term in the perturbation expansion as an electromagnetic mass effect. How could he now get rid of this?

He reasoned that he could just subtract it out. The perturbation expansion for an electron bound in a hydrogen atom includes an infinite mass term. But the expansion for a free electron also includes the same infinite mass term. Why not just subtract one perturbation series from the other, thereby eliminating the infinite terms? It sounds as though subtracting infinity from infinity should yield a nonsensical answer,* but

* Not sure? Try this. The sum of the infinite series of integer numbers, $1 + 2 + 3 + 4 + \cdots$, is obviously infinity. But then, so is the sum of the infinite series of even integer

Bethe now found that in a simple, non-relativistic version of QED this subtraction produced a result that, though it still had problems, behaved in a much more orderly manner. He figured that in a QED that fully complied with Einstein's special theory of relativity, this 'renormalization' procedure would eliminate the problem completely and give a physically realistic answer.

Because the procedure had partly tamed the behaviour of the equations he was able to obtain a rough estimate for the predicted size of the Lamb shift. Uncertain about a factor of 2 he had introduced into the calculation, on getting to the General Electric research laboratories he made a quick visit to the library and reassured himself that he had got it right. He had obtained a prediction for the Lamb shift which was just four per cent larger than the experimental value that Lamb had reported at the Shelter Island conference.

He was definitely on to something.

---

A definitive, relativistic QED that could be renormalized in this way took a little while longer to develop. Schwinger described a version in a marathon, five-hour session delivered at a subsequent conference which took place in March 1948 at the Pocono Manor Inn in the Pocono Mountains near Scranton, Pennsylvania. His mathematics was largely impenetrable.

numbers, $2 + 4 + 6 + 8 + \cdots$ So, let's subtract infinity from infinity by subtracting the series of even numbers from the series of integer numbers. What we get is an infinite series of odd numbers, $1 + 3 + 5 + 7 + \cdots$, which also sums to infinity but which is nevertheless an entirely 'sensible' result. This example is taken from Gribbin, p. 417.

Only Fermi and Bethe, it seemed, followed his derivation through to the end.

Schwinger's New York rival Feynman had in the meantime developed a vastly different, much more intuitive approach to describing and keeping track of the perturbation corrections in QED. Neither understood the other's approach, but when they compared notes at the end of Schwinger's session, they found that their results were identical. 'So I knew that I wasn't crazy,' Feynman said.[2]

This seemed to be the end of the matter, but Oppenheimer learned of yet another successful approach to QED from a letter he received from Japanese physicist Sin-Itiro Tomonaga shortly after returning from the Pocono conference. Tomonaga had used methods similar to Schwinger but his mathematics appeared a lot more straightforward. The situation was rather confusing. These very different approaches to relativistic QED all produced similar answers, but nobody quite understood why.

The challenge was taken up by a young English physicist called Freeman Dyson. On 2 September 1948, he boarded a bus from Berkeley, near San Francisco in California, bound for the East coast. 'On the third day of the journey a remarkable thing happened,' he wrote to his parents a few weeks later. 'Going into a sort of semi-stupor as one does after 48 hours of bus-riding. I began to think very hard about physics, and particularly about the rival radiation theories of Schwinger and Feynman. Gradually my thoughts grew more coherent, and before I knew where I was, I had solved the problem that had been in the back of my mind all this year, which was to prove the equivalence of the two theories.'[3]

The result was a fully relativistic theory of QED that predicts the results of experiments to astonishing levels of accuracy and precision. The *g*-factor for the electron is predicted by QED to have the value 2.00231930476. The comparable experimental value is 2.00231930482.* 'To give you a feeling for the accuracy of these numbers,' Feynman later wrote, 'it comes out something like this: If you were to measure the distance from Los Angeles and New York to this accuracy, it would be exact to the thickness of a human hair.'[4]

---

The success of QED established some important precedents. It now seemed that the correct way to describe a fundamental particle and its interactions was in terms of a quantum field theory in which the force involved is carried by field particles. Like Maxwell's theory of electromagnetism, QED is a U(1) gauge theory in which the local U(1) phase symmetry of the electron wavefunction is connected with the conservation of electric charge.

Attention now turned to a quantum field theory of the strong force between protons and neutrons inside the nucleus. But here was another puzzle. The connection between the conservation of electric charge and electromagnetism – classical or quantum – was intuitively obvious. If a quantum field theory of the strong force was to be discovered, it was first necessary

* These numbers are subject to constant refinement, both experimental and theoretical. The values quoted here are taken from G.D. Coughlan and J.E. Dodd, *The Ideas of Particle Physics: An Introduction for Scientists.* Cambridge University Press, 1991, p. 34.

to figure out what, precisely, was conserved in strong-force interactions and what continuous symmetry transformation this related to.

Chinese physicist Chen Ning Yang believed that the quantity conserved in nuclear interactions involving the strong force was *isospin*.

Yang was born in 1922, in Hefei, the capital city of Anhui province in eastern China. He studied in Kunming, at the National Southwestern Associated University which was formed from Tsinghua, Peking, and Nankai Universities following the invasion of Japanese forces in 1937. He graduated in 1942 and was awarded a master's degree two years later. Armed with a scholarship known as a Boxer Indemnity,* in 1946 he headed for the University of Chicago.

In Chicago he studied nuclear physics under the supervision of Edward Teller. Inspired by reading the autobiography of the American inventor and politician Benjamin Franklin, he adopted a middle name 'Franklin' or just 'Frank'. He obtained his doctorate in 1948 and worked for a further year as an assistant to Fermi. In 1949 he moved to the Institute for Advanced Study in Princeton.

It was in Princeton that he began to think about ways in which he could apply Noether's theorem in search of a quantum field theory of the strong force.

The concept of isospin, or isotopic spin, grew out of the simple fact that the masses of the proton and neutron are very

* This was a scholarship administered by America with funds paid by the Chinese as compensation for the Boxer uprising towards the end of the nineteenth century.

similar.* When the neutron was discovered in 1932, it was natural to assume that this was a composite particle consisting of a proton and an electron. It was well known that beta-radioactive decay involves the ejection of a high-speed electron directly from the nucleus, turning a neutron into a proton in the process. This seemed to imply that in beta-radioactivity, one of the composite neutrons was somehow shedding its 'stuck-on' electron.

Shortly after the discovery of the neutron, Heisenberg used the neutron-as-proton-plus-electron idea to develop an early theory of proton–neutron interactions in the nucleus. This was a model that was closely based on theories of chemical bonding.

Heisenberg hypothesized that the proton and neutron bind together in the nucleus by exchanging an electron between them, the proton turning into a neutron and the neutron turning into a proton in the process. The interaction between two neutrons would involve the exchange of two electrons, one in each 'direction'.

This exchange suggests that in the nucleus, protons and neutrons tend to lose their identity, constantly flitting from one form to another. It suited Heisenberg's purpose to imagine that the proton and neutron are simply different states of the same particle, distinguished by the different properties of

* The masses of sub-atomic particles are typically given as energies, related by Einstein's equation $m = E/c^2$. The proton mass is 938.3 MeV/$c^2$, where MeV means mega (million) electron volts. The neutron mass is 939.6 MeV/$c^2$. The $c^2$ term is often omitted (which means it is implied) and the masses are then given simply as 938.3 and 939.6 MeV, respectively. An electron volt is the amount of energy a single negatively charged electron gains when accelerated through a one-volt electric field.

these states. The different states possess different electrical charges, of course, one positive and one neutral. But to make his theory work he also needed to introduce a further property analogous to electron spin.

He therefore introduced the idea of isospin, not to be confused with electron spin, in which the proton is (arbitrarily) assigned a spin-up orientation and the neutron a spin-down orientation. These are orientations in an 'isospin-space' which has just two dimensions, up and down. Converting a neutron into a proton is then equivalent to 'rotating' the spin of the neutron in isospin space, from spin-down to spin-up.

This all sounds very mysterious, but in many ways isospin is like electrical charge. Our easy familiarity with electricity shouldn't blind us to the fact that this, too, is a property which takes up 'values' (rather than 'orientations') in an abstract 'charge-space' with two dimensions – positive and negative.

Even as a simple analogy, Heisenberg's theory was already a stretch. The strengths of chemical bonds formed by exchanging electrons are much weaker than the strength of the force binding protons and neutrons together inside the nucleus. But Heisenberg was able to use the theory to apply non-relativistic quantum mechanics to the nucleus itself. In a series of papers published in 1932 he accounted for many observations in nuclear physics, such as the relative stabilities of isotopes.

The weaknesses of the theory were exposed in experiments performed just a few years later. Because protons do not possess a 'stuck-on' electron, Heisenberg's electron-exchange model did not allow for any kind of interaction between protons. In contrast, experiments showed that the strength of the interaction between protons is comparable to that between protons and neutrons.

Despite the shortcomings of the theory, Heisenberg's electron-exchange model held at least a grain of truth. The exchange of electrons was abandoned, but the concept of isospin was retained. As far as the strong force is concerned, the proton and the neutron are essentially two states of the same particle, like the two spin orientations of the electron. The only difference between them is their isospin.

---

The individual isospins of protons and neutrons can be added up to produce a total isospin, a concept first introduced by physicist Eugene Wigner in 1937. The literature on nuclear reactions seemed to support the idea that total isospin is conserved, just as electric charge is conserved in physical and chemical changes. Yang now identified isospin as a local gauge symmetry, like the phase symmetry of the electron wavefunction in QED, and began the search for a quantum field theory that would preserve it.

He quickly got bogged down, but became obsessed with the problem. 'Occasionally an obsession does finally turn out to be something good,' he later observed.[5]

In the summer of 1953 he took a short leave of absence from the Institute for Advanced Study and made a visit to Brookhaven National Laboratory on Long Island, New York. He found himself sharing an office with a young American physicist called Robert Mills.

Mills became absorbed by Yang's obsession and together they worked on a quantum field theory of the strong nuclear force. 'There was no other, more immediate motivation,' Mills explained some years later. 'He and I just asked ourselves "Here is something that occurs once. Why not again?"'[6]

In QED, changes in the phase of the electron wavefunction in space and time are compensated by corresponding changes in the electromagnetic field. The field 'pushes back', such that the phase symmetry is preserved. But a new quantum field theory of the strong force had to account for the fact that there are now two particles involved. If isospin symmetry is to be conserved, this means that the strong force sees no difference between the proton and the neutron. Changing the isospin symmetry by, for example, 'rotating' a neutron into a proton, therefore demands a field which 'pushes back' and so restores the symmetry. Yang and Mills therefore introduced a new field, which they called the 'B' field, designed to do just this.

The simple symmetry group U(1) is insufficient for this kind of complexity, and Yang and Mills reached for the symmetry group SU(2), the special unitary group of transformations of two complex variables. A larger symmetry group is needed simply because there are now two objects that can transform into each other.

The theory also needed three new field particles, responsible for carrying the strong force between the protons and neutrons inside the nucleus, analogues of the photon in QED. Two of the three field particles were required to carry electric charge, accounting for the change in charge resulting from proton–neutron and neutron–proton interactions. Yang and Mills referred to these particles as $B^+$ and $B^-$. The third particle was neutral, like the photon, and was meant to account for proton–proton and neutron–neutron interactions in which there is no change in charge. This was referred to as $B^0$. It was found that these field particles interact not only with protons and neutrons, but also with each other.

By the end of the summer they had worked out a solution. But this was a solution with a whole new set of problems.

For one thing, the renormalization methods that had been used so successfully in QED could not be applied to the field theory that Yang and Mills had devised. Worse still, the zeroth-order term in the perturbation expansion indicated that the field particles should be massless, just like the photon. But this was self-contradictory. Heisenberg and Japanese physicist Hideki Yukawa had suggested in 1935 that the field particles of short-range forces like the strong force should be 'heavy', i.e. they should be large, massive particles. Massless field particles for the strong force made no sense whatsoever.

---

Yang returned to Princeton. On 23 February 1954 he presented a seminar on the work he had done with Mills. Oppenheimer was in the audience, as was Pauli, who had moved to Princeton University in 1940.

It turned out that Pauli had earlier explored some of the same logic and had arrived at the same puzzling conclusions concerning the masses of the field particles. He had consequently abandoned the approach. As Yang drew his equations on the blackboard, Pauli piped up:

'What is the mass of this B field?' he asked, anticipating the answer.

> 'I don't know,' Yang replied, somewhat feebly.
> 'What is the *mass* of this B field?' Pauli demanded.
> 'We have investigated that question,' Yang replied. 'It is a very complex question, and we cannot answer it now.'
> 'That is not a sufficient excuse,' Pauli grumbled.[7]

Yang, taken aback, sat down to general embarrassment. 'I think we should let Frank proceed,' Oppenheimer suggested. Yang resumed his lecture. Pauli asked no more questions, but he was irked. The following day he left a note for Yang saying: 'I regret that you made it almost impossible for me to talk with you after the seminar.'[8]

---

It was a problem that simply would not go away. Without mass, the field particles of the Yang–Mills field theory did not fit with physical expectations. If they were massless, as the theory predicted, then they should be as ubiquitous as photons, yet no such particles had ever been observed. The accepted methods of renormalization wouldn't work.

And yet, it was still a *nice* theory.

'The idea was *beautiful* and should be published,' Yang wrote, 'But what is the mass of the gauge particle? We did not have firm conclusions, only frustrating experiences to show that

[this] case is much more involved than electromagnetism. We tended to believe, on physical grounds, that the charged gauge particles cannot be massless.'[9]

Yang and Mills published a paper describing their results in October 1954. In it they wrote: 'We next come to the question of the mass of the [B] quantum, to which we do not have a satisfactory answer.'[10]

They made no further progress, and turned their attentions elsewhere.

# 3

# People Will Be Very Stupid About It

*In which Murray Gell-Mann discovers strangeness and the 'Eightfold Way', Sheldon Glashow applies Yang–Mills field theory to the weak nuclear force, and people are very stupid about it*

Yang and Mills had tried to apply quantum field theory to the problem of strong-force interactions in the hope of repeating the success of QED. But they found that the theory could not be renormalized and yielded massless particles that should have been massive. Obviously, this could not be the solution to the strong force.

But what of the weak nuclear force?

The weak force was something of a mystery. Italian physicist Enrico Fermi had been obliged to invoke a new type of nuclear force in a detailed theory of beta-radioactivity in the early 1930s. He described this theory to his colleagues during a group skiing holiday in the Italian Alps at Christmas 1933. His colleague Emilio Segrè subsequently described the experience: '. . . we were all sitting on one bed in a hotelroom, and I could hardly keep still in that position, bruised as I was after several falls on icy snow. Fermi was fully aware of the importance of his accomplishment and said that he thought he would be remembered for this paper, his best so far.'[1]

Fermi drew parallels between the weak force and electromagnetism. From the resulting electromagnetic-like theory, he was able to deduce the range of energies (and hence speeds) of the emitted beta-electrons. His predictions were shown to be correct in experiments performed at Columbia University by Chinese–American physicist Chien-Shiung Wu in 1949. With some small adjustments, Fermi's theory remains valid to this day.

Fermi deduced that the strength of the interactions between the particles involved in beta-radioactivity is some ten billion times weaker than electromagnetic interactions between charged particles. This is weak indeed, but the force has some profound consequences. Because of the weak force, neutrons are inherently unstable. A neutron moving in free space will survive intact for an average of just 18 minutes. This is unusual behaviour for a particle that is meant to be fundamental or elementary.*

Of course, it was a bit much to have to invoke a novel force of nature just to explain a single type of interaction. But as experimentalists began to sift through the 'zoo' of new particles

* Those looking for an even more profound consequence of weak-force interactions should look no further than the Standard Solar Model, the contemporary theory describing how the sun works. The fusion of protons (hydrogen nuclei) to form helium nuclei at the sun's core involves transformation of two protons into two neutrons via the weak force, accompanied by the emission of two positrons and two neutrinos.

that were now being revealed in the debris of high-energy collisions, evidence began to emerge for other kinds of particles that were susceptible to the weak force.

---

In the 1930s, if you wanted to study high-energy particle collisions then you needed to climb a mountain. Cosmic rays – streams of high-energy particles from outer space – wash constantly over the upper atmosphere. Some very highly energetic particles that constitute these rays can penetrate to lower levels of the atmosphere, levels that can be reached from the tops of mountains, where their collisions can be studied. Such studies rely on chance detection of the particles and, because of their randomness, no two events ever have quite the same conditions.

American physicist Carl Anderson had discovered Dirac's positron in 1932. Four years later, he and fellow American Seth Neddermeyer loaded their particle detection apparatus onto a flat-bed truck and drove to the top of Pike's Peak (now Pikes Peak) in the Rocky Mountains, about ten miles west of Colorado Springs.* In the tracks left by penetrating cosmic rays, the physicists identified another new particle. This particle behaved just like an electron, but was found to be deflected by a magnetic field to a much lesser extent.

* In fact, their truck couldn't quite make it all the way to the toll gate and they had to be towed the rest of the way. The scientists' budget for these experiments was extremely limited but they were fortunate to encounter a vice-president of General Motors, testing a new Chevrolet truck on the mountain. He kindly arranged for the scientists' truck to be towed and paid for the engine to be replaced.

The particle curved more slowly than an electron, and more sharply than a proton of similar speed would curve (in the opposite direction). There was no alternative but to conclude that this was a new 'heavy' electron, with a mass about 200 times that of an ordinary electron. It could not be a proton, as the proton mass is about 2000 times that of an electron.*

The new particle was initially called the mesotron, subsequently shortened to meson. It was an unwelcome discovery. A heavy version of the electron? This did not fit with any theories or preconceptions of how the fundamental building blocks of nature should be organized.

Incensed, Galician-born American physicist Isidor Rabi demanded to know: 'Who ordered that?'[2] Willis Lamb echoed this sense of frustration in his 1955 Nobel lecture when he said: '... the finder of a new elementary particle used to be rewarded by a Nobel Prize, but such a discovery now ought to be punished by a $10,000 fine.'[3]

In 1947 another new particle was discovered in cosmic rays atop the Pic du Midi in the French Pyrenees by Bristol University physicist Cecil Powell and his team. The new particle was found to have a slightly larger mass than the meson; 273 times that of the electron. It came in positive, negative, and, subsequently, neutral varieties.

The physicists were now running into trouble with names. The meson was renamed the mu-meson, subsequently

* Actually, the ratio of the proton and electron rest masses (the masses that these particles would possess at zero speed) is 1836.

shortened to muon.* The new particle was called the pi-meson (pion). As techniques for detecting particles produced by cosmic rays became more sophisticated, the floodgates opened. The pion was quickly followed by the positive and negative K-meson (kaon) and the neutral lambda particle. New names proliferated. Responding to a question from one young physicist, Fermi remarked: 'Young man, if I could remember the names of these particles, I would have been a botanist.'[4]

The kaons and the lambda behaved rather oddly. These particles were produced in abundance, a signature of strong-force interactions. They were often produced in pairs which formed characteristic 'V'-shaped tracks. They would then travel on through the detector before disintegrating. Their disintegration took a lot longer than their production, suggesting that although they were being produced by the strong force, their decay modes were governed by a much weaker force, the same force, in fact, that governs radioactive beta-decay.

Isospin could not help to explain the strange behaviour of the kaons and the lambda. It seemed as if these new particles possessed some additional, hitherto unknown, property.

American physicist Murray Gell-Mann was puzzled. He realized that he could account for the behaviour of these new particles using isospin provided he assumed that the isospins were for some reason 'shifted' by one unit. This

* This was a confusing time. As will become apparent shortly, the mu-meson does not in fact belong to the class of particles that would collectively become known as 'mesons'.

made no sense physically, so he proposed a new property, which he subsequently called *strangeness,* to account for this shift.* He later immortalized the term with the words of Francis Bacon: 'There is no excellent beauty that hath not some strangeness in the proportion.'[5]

Gell-Mann argued that, whatever it is, strangeness is, like isospin, conserved in strong-force interactions. In a strong-force interaction involving an 'ordinary' (i.e. non-strange) particle, production of a strange particle with a strangeness value of +1 had to be accompanied by another strange particle with a strangeness value of –1, so that the total strangeness was conserved. This was why the particles tended to be produced in pairs.

Conservation of strangeness also explained why the strange particles took so long to decay. Once formed, transformation of each strange particle back into ordinary particles was not possible through strong-force interactions, which could be expected to happen quickly, because this would require a change in strangeness (from +1 or –1 to zero). The strange particles therefore hung around long enough to succumb to the weak force, which does not respect the conservation of strangeness.

Nobody knew why.

---

* Much the same idea was put forward at around the same time by Japanese physicists Kazuhiko Nishijima and Tadao Nakano, who referred to strangeness as 'η-charge'.

In his landmark paper on beta-radioactivity, Fermi had drawn an analogy between the weak force and electromagnetism. He had made his estimates of the relative strengths of the forces involved in the interaction using the mass of the electron as a yardstick. In 1941, Julian Schwinger had wondered what the consequences would be if he assumed that the weak force is carried by a much, much larger particle. He had estimated that if this field particle was actually a couple of hundred times the proton mass, then the strengths of weak-force and electromagnetic interactions might actually be the same. This was the first hint that it might be possible to *unify* the weak and electromagnetic forces into a single, 'electro-weak' force.

Yang and Mills had discovered that, to take account of all the different ways that neutrons and protons can interact in the nucleus, they needed three different kinds of force particles. In 1957, Schwinger came to much the same conclusion regarding weak-force interactions. He published an article in which he speculated that the weak force is carried by three field particles. Two of these particles, the $W^+$ and $W^-$ (in modern parlance) are necessary to account for the transmission of electric charge in weak interactions. A third, neutral particle is needed to account for instances in which no charge is transferred. Schwinger believed that this third particle was the photon.

According to Schwinger's scheme beta-radioactivity would now work like this. A neutron would decay, emitting a massive $W^-$ particle and turning into a proton. The short-lived $W^-$ particle would in its turn decay into a high-speed electron (the beta-particle) and an anti-neutrino (see Figure 8).

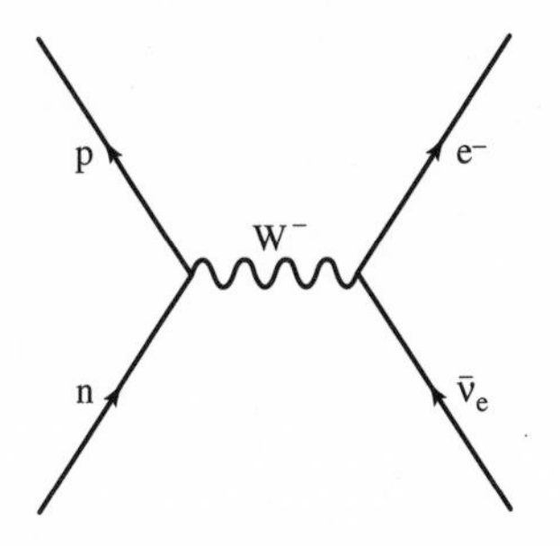


FIGURE 8 The mechanism of nuclear beta-decay could now be explained in terms of the decay of a neutron (n) into a proton (p), with the emission of a virtual $W^-$ particle. The $W^-$ particle goes on to decay into an electron ($e^-$) and an anti-neutrino ($\bar{\nu}_e$).

Schwinger asked one of his Harvard graduate students to work on the problem.

Sheldon Glashow was an American-born son of Russian Jewish immigrants. He graduated from the Bronx High School of Science in 1950 together with his classmate Steven Weinberg. He had gone with Weinberg to Cornell University, securing his bachelor's degree in 1954, before moving on to become one of Schwinger's graduate students at Harvard.

The heavy W particles that Schwinger had hypothesized were obliged to carry electric charge. Glashow soon realized that this simple fact meant that it was actually impossible to separate the theory of the weak force from that of electromagnetism. 'We should care to suggest,' he wrote in an appendix to his PhD thesis, 'that a fully acceptable theory of these interactions may only be achieved if they are treated together . . .'[6]

Glashow now reached for the same SU(2) quantum field theory that Yang and Mills had developed, taking on faith Schwinger's assertion that the three field particles of the weak force were the two heavy W particles and the photon. For a time he believed he had succeeded in developing a unified theory of the weak and electromagnetic forces. What's more, he believed that his theory was renormalizable.

But in truth he had made a series of errors. When these were revealed, he realized that the theory was demanding too much from the photon. His solution was to enlarge the symmetry by combining the Yang–Mills SU(2) gauge field with the U(1) gauge field of electromagnetism, in a product written SU(2) × U(1). This represents more of a 'mixture' of weak and electromagnetic forces rather than a fully unified electro-weak force, but it had the advantage that it freed the photon from its burden of responsibility for aspects of weak-force interactions.

The theory still demanded a neutral carrier for the weak force. Glashow now had three massive weak-force particles equivalent to the triplet of B particles first introduced by Yang and Mills. These were the $W^+$, $W^-$, and $Z^0$.*

In March 1960 Glashow lectured in Paris. Here he encountered Gell-Mann, on sabbatical leave from the California Institute of Technology (Caltech) and working as a visiting professor

* Glashow originally referred to the neutral particle as B, by analogy with Yang and Mills, but it is now commonly referred to as the $Z^0$.

at the Collège de France. Glashow described his SU(2) × U(1) theory over lunch. Gell-Mann offered encouragement. 'What you're doing is good,' Gell-Mann told him, 'But people will be very stupid about it.'[7]

Stupid or not, the physics community was largely unimpressed with Glashow's theory. Just as Yang and Mills had discovered, the SU(2)×U(1) field theory predicted that the carriers of the weak force should be massless, like the photon. Inserting the masses 'by hand' into the equations would ensure that the theory remained unrenormalizable. Like Yang and Mills before him, Glashow could not figure out how the field particles were supposed to acquire their mass.

There was more trouble. Elementary-particle interactions involve one or more particles decaying or reacting together to produce new particles. When such interactions involve charged intermediaries their reactions are referred to as charged 'currents', as they involve the 'flow' of charge from the starting to the finishing particles. It was anticipated that a neutral weak-force carrier – the $Z^0$ – would manifest itself experimentally in the form of interactions involving no change in charge, called 'neutral currents'. No evidence for any such currents could be found in the strange-particle decays, which by now had become the particle physicists' principal hunting ground for data on weak-force interactions.

Glashow waved his arms. He argued that the $Z^0$ was simply so much more massive than the charged W particles that interactions involving the $Z^0$ were out of reach of contemporary experiments. The experimentalists were not impressed.

---

Murray Gell-Mann had been born in New York in 1929. A child prodigy, he entered Yale University to study for a bachelor's degree when he was just fifteen. He secured his doctorate at the Massachusetts Institute of Technology (MIT) in 1951, aged just 21. He worked for a short time at the Institute for Advanced Study in Princeton before moving first to the University of Illinois at Urbana-Champaign, then Columbia University in New York, then the University of Chicago where he worked with Fermi and puzzled over the properties of the strange particles.

In 1955 he took a professorship at Caltech where he worked with Feynman on the theory of the weak nuclear force. He also began to turn his attention to the problem of classifying the 'zoo' of elementary particles that had by now been discovered. It was possible to discern small patterns in the zoo – particles that clearly belonged to the same species, for example – but the individual patterns did not fit together to give a coherent picture.

Particle physicists had by this time introduced a taxonomy to lend the zoo at least some sense of order. There were two principal classes. These were the *hadrons* (from the Greek *hadros*, meaning thick or heavy) and the *leptons* (from the Greek *leptos*, meaning small).

The class of hadrons includes a sub-class of *baryons* (from the Greek *barys*, also meaning heavy). These are heavier particles which experience the strong nuclear force and include the proton (p), neutron (n), lambda ($\Lambda^0$), and two further series of particles that had been discovered in the 1950s and named sigma ($\Sigma^+$, $\Sigma^o$, and $\Sigma^-$) and xi ($\Xi^o$, $\Xi^-$). The class of hadrons also includes the sub-class of *mesons* (from the Greek *mésos*), meaning 'middle'). These particles experience the strong force but are of intermediate mass, such as the pions ($\pi^+$, $\pi^o$, $\pi^-$) and the kaons ($K^+$, $K^o$, and $K^-$).

The class of leptons includes the electron ($e^-$), muon ($\mu^-$), and the neutrino ($\nu$). These are light particles which do not experience the strong nuclear force. Both the baryons and the leptons are *fermions,* named for Enrico Fermi. They are characterized by half-integral spins. The baryons and leptons listed above all possess a spin of ½, and can therefore take up two spin orientations, given as +½ (spin-up) and –½ (spin-down). Fermions obey Pauli's exclusion principle.

Sitting outside the classes of hadrons and leptons was the photon, the carrier of the electromagnetic force. The photon is a *boson,* named for Indian physicist Satyendra Nath Bose. Bosons are characterized by integral spin quantum numbers and are not subject to Pauli's exclusion principle. Other force carriers, such as the hypothetical $W^+$, $W^-$, and $Z^0$ particles, were expected to be bosons with integral spins. Bosons with zero spins are also possible, but these are not force particles. Mesons are examples of bosons with zero spin. The classification of particles known around 1960 is summarized in Figure 9.

It was clear that amidst this confusion there must be a pattern, a particle equivalent of Dmitri Mendeleev's periodic table of the elements. The question was: What is this pattern and does it have an underlying explanation?

Gell-Mann initially tried to construct a pattern from a fundamental triplet of particles consisting of the proton, neutron, and lambda, using these as building blocks to construct all the other

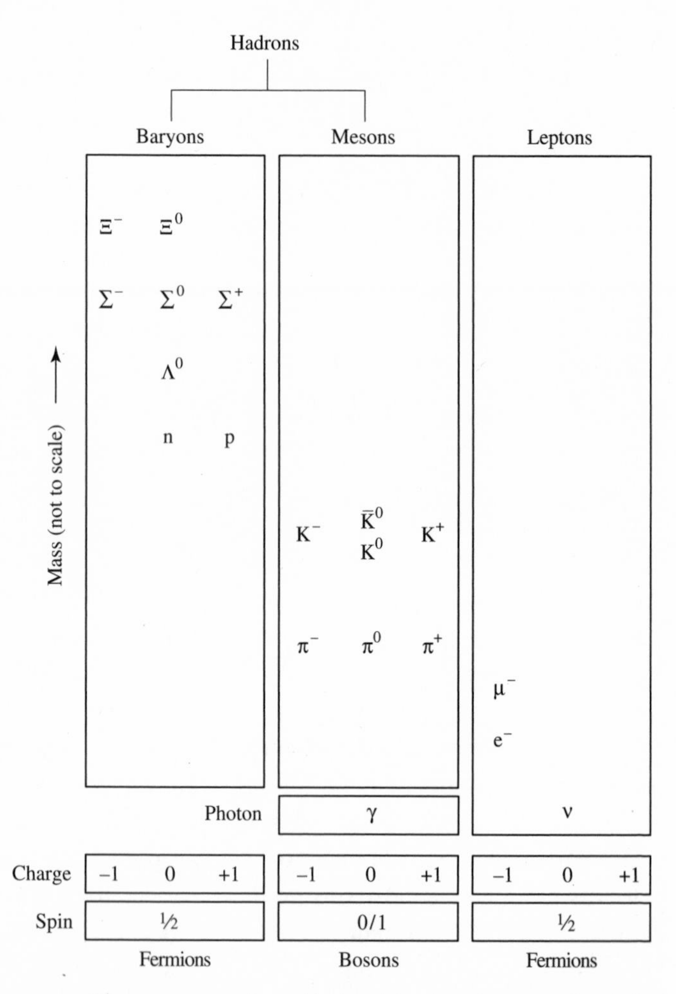


**FIGURE 9** The taxonomy adopted by particle physicists around 1960 had helped to organise the known particles into different classes. These were hadrons (baryons and mesons) and leptons. Sitting outside this classification was the photon, the force particle of electromagnetism.

hadrons. But it was a big mess. It was never really clear why these particles should be regarded as more 'fundamental' than the others. He realized that he was reaching for the underlying explanation before a proper pattern had been established. This was a bit like trying to figure out the fundamental building blocks of the chemical elements without first appreciating the position that each element occupies in the periodic table.

Gell-Mann believed that the framework for such a pattern could be provided by a global symmetry group, a way of organizing the particles so that the pattern of their interrelationships could be revealed. He was at this stage searching only for a way of reclassifying the particles, rather than seeking to develop a Yang–Mills field theory, which would have required a local symmetry.

He knew he needed a larger continuous symmetry group than U(1) or SU(2) to accommodate the range and variety of particles that were then known, but he was quite unsure how to proceed. By this time he was working as a visiting professor at the Collège de France in Paris. Perhaps not surprisingly, copious quantities of good French wine consumed over lunch with his French colleagues did not immediately help to point the way to a solution.

Glashow's visit to Paris in March 1960 therefore prompted more than just noises of encouragement. Gell-Mann was intrigued by Glashow's SU(2) × U(1) theory. He began to understand how it might be possible to expand the symmetry group to higher dimensions. Thus inspired, he now tried theories with more and more dimensions. He tried three, four, five, six, and seven dimensions, trying to find a structure that did not correspond to the product of SU(2) and U(1).

'At that point, I said, 'That's enough!' I did not have the strength after drinking all that wine to try eight dimensions.'[8]

It seemed that the wine had not aided conversation either. The colleagues with whom Gell-Mann was drinking at lunch were mathematicians who could have solved his problem almost immediately. But he never discussed it with them.

Glashow chose to accept Gell-Mann's offer to join him at Caltech and, shortly after his return from Paris, the two physicists searched for a solution together. But it was only after a chance discussion with Caltech mathematician Richard Block that Gell-Mann discovered that the Lie group SU(3) offered the structure he had been searching for. In Paris he had given up just as he was about to discover this for himself.

The simplest, or so-called 'irreducible' representation of SU(3) is a fundamental triplet. Other theorists had actually tried to construct a model based on the SU(3) symmetry group and had used the proton, neutron, and lambda particles as the fundamental representation. Gell-Mann had already been down this road, and had no wish to repeat his experiences. He simply skipped over the fundamental representation and turned his attention to the next.

One of the representations of SU(3) consists of eight dimensions. 'Rotating' a particle in one dimension transforms it into a particle in another dimension, just as 'rotating' the isospin of the neutron in the SU(2) symmetry group turns it into a proton. If Gell-Mann could somehow place a particle in each dimension, then perhaps he could begin to understand the nature of their underlying relationships. It was surely no coincidence that there were eight baryons – the proton, neutron, lambda, three sigma, and two xi particles?

These particles could be distinguished by their values of electric charge, isospin, and strangeness. Plot strangeness value against either charge or isospin on a graph and a hexagonal pattern emerges with a particle at each apex and two particles in the centre (see Figure 10). The pattern demanded that the proton, neutron, and lambda particles be included in the scheme, and Gell-Mann must have felt justified in his decision to resist ascribing these to the fundamental representation.

When Gell-Mann produced a similar analysis for the mesons, he found that he needed to include the anti-$K^o$ but was still short by one particle. The meson equivalent of the

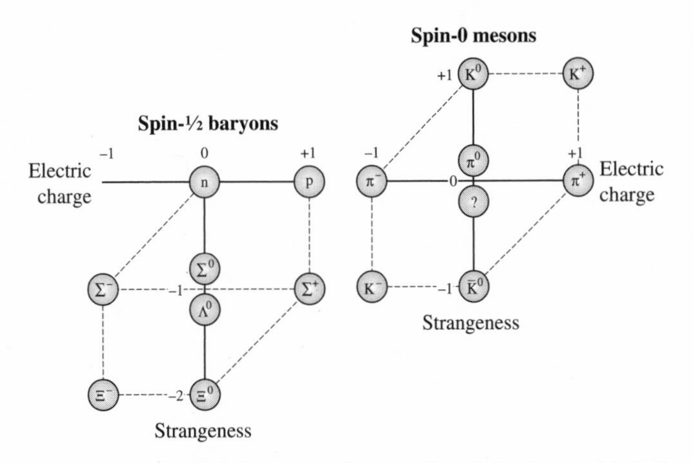


**FIGURE 10** The Eightfold Way. Gell-Mann found that he could fit the baryons, including the neutron (n), and proton (p) and the mesons into two octet representations of the global symmetry group SU(3). But there were only seven particles in the representation for the mesons. One particle, the meson equivalent of the $\Lambda^o$, was missing. This particle was found a few months later by Luis Alvarez and his team in Berkeley. They called it the eta, $\eta$.

lambda was 'missing'. Emboldened, he speculated that there must exist an eighth meson with an electric charge of zero and zero strangeness.

Gell-Mann had discovered the patterns in two 'octets' of particles based on an eight-dimensional representation of the global SU(3) symmetry group. He called it the 'Eightfold Way', a tongue-in-cheek reference to the teachings of Buddha on the eight steps to Nirvana.* He completed his work on the Eightfold Way during Christmas 1960 and it was published as a Caltech preprint in early 1961. The particle he had predicted to complete the meson octet was found a few months later by American physicist Luis Alvarez and his team in Berkeley, California. They called the new particle the eta, η.

---

Gell-Mann was working alone but he was not the only theorist searching for a pattern. Yuval Ne'eman was a late entrant to the firmament of theoretical physics. Where Gell-Mann had gone to Yale at the tender age of fifteen, Ne'eman, a native of Tel Aviv, had joined the Haganah, the Jewish underground, in what was then the British Mandate for Palestine. He had commanded an infantry battalion in the 1948 Arab–Israeli war and served as acting head of the Israeli Secret Service.

He had achieved the rank of Colonel in the Israeli Defense Force when he decided to seek an opportunity to study for a doctorate in physics. Moshe Dayan, defense chief of staff, agreed to appoint him as a defence attaché at the Israeli

* These are: right views, right intention, right speech, right action, right living, right effort, right mindfulness, and right concentration.

Embassy in London. Dayan figured that Ne'eman could study for his PhD in his spare time.

Ne'eman had originally intended to study relativity at King's College in London, but he quickly discovered that the city traffic made it impossible for him to get there from the Embassy in Kensington in time to attend lectures and seminars. He switched to Imperial College and particle physics. At Imperial College he was pointed in the direction of Pakistan-born theorist Abdus Salam.

Ne'eman worked in the evenings and at weekends. He began a search for symmetry groups that might accommodate the known particles and turned up five candidates, including SU(3). Initially excited by the very resonant possibilities afforded by a symmetry group that produced a Star of David pattern, Ne'eman eventually fixed on SU(3). He published his own version of the Eightfold Way in July 1961.

Salam was initially sceptical, but when a draft of Gell-Mann's paper arrived on his desk he quickly set aside his reservations. Despite having a slight head start, Ne'eman was beaten into print by Gell-Mann (although Ne'eman's paper was actually the first to be published in a physics journal). But he was not disappointed. On the contrary, he was thrilled to find himself in such good company.

Both Ne'eman and Gell-Mann attended a particle physics conference in June 1962, held at the Organisation Européenne pour la Recherche Nucléaire (CERN) in Geneva. Both listened intently to reports of further new particles that had been discovered, a triplet of what later came to be called sigma-star particles with strangeness values of −1, and a doublet of xi-star particles with strangeness values of −2.

Ne'eman saw immediately that these particles belonged to another representation of SU(3) consisting of ten dimensions. It took him just a moment to realize that of the ten particles implied by this representation, nine had now been found. The particle needed to complete the pattern was negatively charged with a strangeness value of −3.

He raised his hand to speak, but Gell-Mann had made precisely the same connection and was sitting closer to the front of the auditorium. It was therefore Gell-Mann who stood to predict the existence of a particle he called the omega. It was discovered in January 1964.

The pattern had now been found, but what of the underlying explanation?

# 4

# Applying the Right Ideas to the Wrong Problem

*In which Murray Gell-Mann and George Zweig invent quarks and Steven Weinberg and Abdus Salam use the Higgs mechanism to give mass to the W and Z particles (finally!)*

Japanese-born American physicist Yoichiro Nambu was deeply worried.

Nambu had studied physics at Tokyo Imperial University, graduating in 1942. He was drawn to particle physics by the reputations of Yoshio Nishina, Sin-Itiro Tomonaga, and Hideki Yukawa, the founders of particle physics in Japan. But there was no great particle physicist in Tokyo, so he studied the physics of solids instead.

From Tokyo, Nambu moved in 1949 to take up a professorship at Osaka City University. Three years later he was invited to the Institute for Advanced Study in Princeton. He moved to the University of Chicago in 1954 and was appointed to a professorship there four years later.

In 1956 he attended a seminar given by John Schrieffer on the new theory of superconductivity he had developed together with John Bardeen and Leon Cooper. This was an elegant application of quantum theory to explain why certain crystalline materials, when cooled below a critical temperature, lose all their electrical resistance. They become superconductors.

Like charges repel each other. However, electrons in a superconductor experience a weak mutual *attraction*. What happens is that a free electron passing close to a positively charged ion in the crystal lattice exerts an attractive force which pulls the ion out of position slightly, distorting the lattice. The electron moves on, but the distorted lattice continues to vibrate back and forth. This vibration produces a slight excess positive charge, which attracts a second electron.

The upshot of this interaction is that a pair of electrons (called a 'Cooper pair'), each with opposite spin and momentum, move through the lattice cooperatively, their motion mediated or facilitated by the lattice vibrations. Recall that electrons are fermions and, as such, they are forbidden from occupying the same quantum state by Pauli's exclusion principle. In contrast, Cooper pairs behave like bosons, which are not so constrained. There is no restriction on the number of pairs that can occupy a quantum state and at low temperatures they can 'condense', gathering in a single state which can build to macroscopic dimensions.* The Cooper pairs in this state experience no resistance as they pass through the lattice and the result is superconductivity.

What worried Nambu was that the theory did not seem to respect the gauge invariance of the electromagnetic field. In

* Laser light is an example of this kind of condensation involving photons.

other words, it did not seem to respect the conservation of electric charge.

Nambu nagged away at this problem and was able to draw on his background in solid-state physics. He realized that the Bardeen–Cooper–Schrieffer (BCS) theory of superconductivity is an example of *spontaneous symmetry-breaking* applied to the gauge field of electromagnetism.

Examples of symmetry-breaking are common. A pencil balanced on its tip is perfectly symmetrical, but very unstable. When it topples over it does so in a specific (though apparently random) direction, and the symmetry is said to be spontaneously broken. Likewise, a marble balanced atop a sombrero is perfectly symmetrical, but unstable. The marble rolls down a specific (though apparently random) direction and comes to rest in the shallow brim of the hat. In truth, tiny fluctuations of the background environment are responsible for the pencil toppling over or the marble rolling down the hat. These tiny fluctuations form part of the background 'noise'.

Spontaneous symmetry-breaking affects the lowest-energy, so-called 'vacuum' state of a system. Like any material, a superconductor could be expected to have a vacuum state in which all particles retain fixed positions in the lattice structure and its electrons remain motionless. However, the possibility of cooperative motions of Cooper pairs mediated by lattice vibrations results in a vacuum state that is *lower* in energy. In this case, the U(1) gauge symmetry of electromagnetism is broken by the presence of another quantum field, whose quanta are the Cooper pairs. The laws describing the dynamics of electrons in the material remain invariant under the local U(1) gauge symmetry, but the vacuum state does not.

Nambu realized that because the Cooper pairs exist in a state of lower energy, it is now necessary to input energy to break them apart. Free electrons created in this way would possess an additional energy equal to half the energy required to break the pairs apart. This additional energy would appear as extra mass. He was struck by the possibilities, and summarized these some years later as follows:[1]

> What would happen if a kind of superconducting material occupied all of the universe, and we were living in it? Since we cannot observe the true vacuum, the [lowest-energy] ground state of this medium would become the vacuum, in fact. Then even particles which were massless . . . in the true vacuum would acquire mass in the real world.

Break the symmetry, Nambu reasoned, and you get particles with mass.

In 1961, Nambu and Italian physicist Giovanni Jona-Lasinio published a paper which outlined just such a mechanism. To get it to work, they had to invoke a background quantum field which creates a 'false' vacuum. In the above example, the pencil topples over when it interacts with the background 'noise', breaking the symmetry. Similarly, to break the symmetry in a quantum field theory requires a background with which to interact. What this implies is that empty space is not actually empty. It contains energy in the form of an all-pervasive quantum field.

In their model, this false vacuum provided the background required to break the symmetry in a theory of strong-force interactions involving hypothetical massless protons and neutrons. The result was indeed protons and neutrons with mass. Breaking the symmetry had 'switched on' the particle masses.

But this was not plain-sailing. British-born physicist Jeffrey Goldstone also studied symmetry-breaking and concluded that one consequence is the creation of yet another massless particle.

In fact, Nambu and Jona-Lasinio had actually stumbled across the same problem in their model. In addition to giving mass to protons and neutrons, their model also predicted massless particles formed from nucleons and anti-nucleons. In their paper they had tried to argue that these may actually acquire a small mass and so could be identified as pions.

These new massless particles came to be called *Nambu–Goldstone bosons*. Goldstone felt instinctively that the creation of these particles would prove to be a general result, applicable for all symmetries, and in 1961 elevated it to the status of a principle. It became known as the *Goldstone theorem*.

Of course, these Nambu–Goldstone bosons suffered from precisely the same objections as the massless particles of the quantum field theories. Any new massless particles predicted by theory could be expected to be as ubiquitous as photons. But, of course, these additional particles had never been observed.

Spontaneous symmetry-breaking promised a solution to the problem of massless particles in Yang–Mills field theories. Yet symmetry-breaking had to be accompanied by yet more massless particles that had never been seen. As one problem was fixed, another was created. If any progress was to be made, some way of avoiding or beating the Goldstone theorem had to be found.

---

Both Gell-Mann and Ne'eman had skipped over the fundamental representation of the global SU(3) symmetry group. They had found that they could accommodate the proton and neutron in the next, eight-dimensional representation, applied to baryons. The implications were fairly obvious. The eight members of the baryon octet – including the proton and neutron – must be composites formed from three even more fundamental particles unknown to experimental science. Obvious, perhaps, but this was a conjecture with some very uncomfortable consequences.

In 1963 Robert Serber at Columbia University began to toy with combinations of three (unspecified) fundamental particles to create the two octets of the Eightfold Way. In this model, each member of the baryon octet would be formed from combinations of the three new particles, and the meson octet from combinations of the fundamental particles and their anti-particles. When in March that year Gell-Mann arrived at Columbia University to deliver a series of lectures, Serber asked him what he thought about this idea.

The conversation took place over lunch at the Columbia Faculty Club.

'I pointed out that you could take three pieces and make protons and neutrons,' Serber explained. 'Pieces and anti-pieces could make mesons. So I said "Why don't you consider that?"'[2]

Gell-Mann was dismissive. He asked Serber what the electric charges of this new triplet of fundamental particles would need to be, something Serber hadn't considered.

'It was a crazy idea,' Gell-Mann said. 'I grabbed the back of a napkin and did the necessary calculations to show that to do this would mean that the particles would have to have fractional electric charges – $-\frac{1}{3}$, $+\frac{2}{3}$, like so – in order to add up to a proton or neutron with a charge of plus one or zero.'[3]

Serber agreed that this was an appalling result. Just twelve years after the discovery of the electron, American physicists Robert Millikan and Harvey Fletcher had performed their famous 'oil drop' experiment, measuring the fundamental unit of electric charge carried by a single electron. When reported in standard units, the charge on the electron is a complicated number with many decimal places,* but it was quickly recognized that all charged particles carry charges that are integral multiples of this fundamental unit. At no time in the 54 years that had elapsed since the notion of a fundamental unit of charge had been established had there been even the merest hint that there might exist particles with charge less than this.

In their subsequent discussion Gell-Mann called Serber's new particles 'quorks', a nonsense word deliberately chosen to highlight the absurdity of the suggestion. Serber took the word as a derivative of 'quirk', as Gell-Mann had said that such particles would indeed be a strange quirk of nature.

* The currently accepted value of the charge on the electron is 1.602176487 (40)$\times 10^{-19}$ coulomb, where the numbers in brackets represent the uncertainty in the last two decimal places.

But despite the appalling consequences, the logic was inescapable. The SU(3) symmetry group demanded a fundamental representation and the fact that the known particles could be fitted into the two octet patterns was very suggestive of a triplet of fundamental particles. The fractional charges were problematic but perhaps, Gell-Mann now reasoned, if the 'quorks' were forever trapped or *confined* inside the larger hadrons then this might explain why fractionally charged particles had never been seen in experiments.

As Gell-Mann's ideas took shape, he happened on a passage from James Joyce's *Finnegan's Wake* which gave him a basis for the name of these ridiculous new particles:

> Three quarks for Muster Mark!
> Sure he hasn't got much of a bark.
> And sure any he has it's all beside the mark.

'That's it!' he declared, 'Three quarks make a neutron and a proton!' The word didn't quite rhyme with his original 'quork' but it was close enough. 'So that was the name I chose. The whole thing is just a gag. It's a reaction against pretentious scientific language.'[4]

Gell-Mann published a two-page article explaining this idea in February 1964. He referred to the three quarks as u, d, and s. Although he didn't say so in his paper, these stood for 'up' (u), with a charge of $+\frac{2}{3}$, 'down' (d) with a charge of $-\frac{1}{3}$, and 'strange' (s), also with a charge of $-\frac{1}{3}$. Baryons are formed from various permutations of these three quarks, and mesons from combinations of quarks and anti-quarks.

In this scheme the proton consists of two up-quarks and a down-quark (uud), with a total charge of +1. The neutron consists of an up-quark and two down-quarks (udd), with a total charge of zero. As the model was elaborated, it transpired that isospin is related to the content of up- and down-quarks in the composite particle. The neutron and proton possess isospins that can be calculated as half the number of up-quarks minus the number of down-quarks.* For the neutron this gives an isospin of $\frac{1}{2} \times (1 - 2)$, or $-\frac{1}{2}$. 'Rotating' the isospin of the neutron is then equivalent to changing a down-quark into an up-quark, giving a proton with an isospin of $\frac{1}{2} \times (2 - 1)$, or $+\frac{1}{2}$. The conservation of isospin now becomes the conservation of quark number. Beta-radioactivity now involves the conversion of a down-quark in a neutron into an up-quark, turning the neutron into a proton, with the emission of a $W^-$ particle, as shown in Figure 11.

The 'strange' particles have strangeness values given simply as minus the number of strange-quarks present.† It was now apparent that a graph of charge or isospin versus strangeness simply maps out the quark content of the particles, with different combinations of quarks appearing at different locations on the map (see Figure 12).

Once again, Gell-Mann was working alone but was not the only theorist on the trail of an underlying explanation. Having returned from Britain to Israel a couple of years before,

* The relation is a little bit more involved than this. In fact, the isospin is given as half×(number of up-quarks minus number of anti-up-quarks) minus (number of down-quarks minus number of anti-down-quarks).

† Again, the relation is a bit more involved. Strangeness is given as minus (number of strange-quarks minus number of anti-strange-quarks).

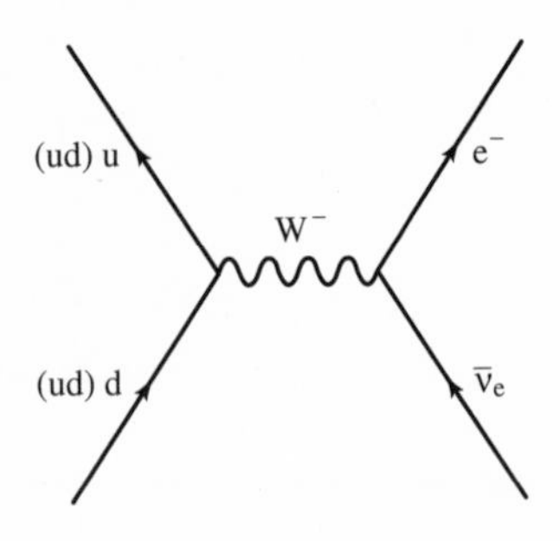


**FIGURE 11** The mechanism of nuclear beta-decay is now explained in terms of the weak-force decay of a down-quark inside a neutron (d) into an up-quark (u), turning the neutron into a proton, with the emission of a virtual $W^-$ particle.

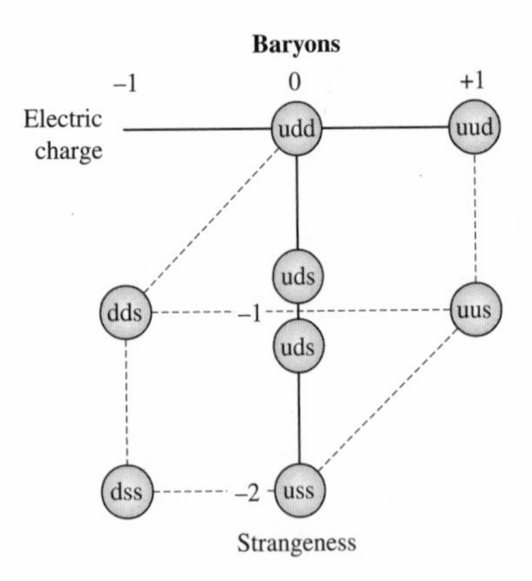


**FIGURE 12** The Eightfold Way could be neatly explained in terms of the various possible combinations of up-, down- and strange-quarks, illustrated here for the baryon octet. The $\Lambda^o$ and $\Sigma^0$ are both composed of up-, down-, and strange-quarks but differ in their isospin. The $\Lambda^o$ has isospin zero and $\Sigma^0$ has isospin 1. This difference can be traced to the different possible combinations of the up-down quark wavefunctions. The $\Lambda^o$ has an anti-symmetric (ud – du) combination, the $\Sigma^0$ has a symmetric (ud + du) combination.

Ne'eman and Israeli mathematician Haim Goldberg had worked on a very speculative proposal concerning a fundamental triplet, but they had stepped back from declaring that these could be 'real' particles with fractional electric charges.

At around the same time that Gell-Mann's speculations appeared in print, former Caltech student George Zweig had developed an entirely equivalent scheme based on a fundamental triplet of particles that he called 'aces'. He figured that baryons could be constructed from 'treys' (triplets) of aces and the mesons from 'deuces' (doublets) of aces and anti-aces. Zweig was working as a postdoctoral associate at CERN, and published his ideas as a CERN preprint in January 1964. Having subsequently seen Gell-Mann's paper, he moved quickly to elaborate the model, produced a second, 80-page CERN preprint, and submitted this to the prestigious journal *Physical Review*.

He was shouted down by his peer reviewers. The paper was never published.

Gell-Mann was already an established physicist, with many notable discoveries to his credit, and could be forgiven his 'lapse' of judgement over the quarks. As a young postdoctoral associate, Zweig was not in such a fortunate position. When shortly afterwards he sought an appointment at a leading university, one of the faculty members, a respected senior theorist, declared the ace model to be the work of a charlatan. Zweig was denied the appointment, and re-joined the Caltech faculty in late 1964. Gell-Mann later took pains to ensure Zweig was credited with his role in the discovery of quarks.

The quark model was a beautifully simplifying scheme, but in truth it was not much more than the result of playing with the patterns. There was simply no experimental foundation for it. Gell-Mann didn't help his cause by being rather cagey about the status of the new particles. Wishing to avoid getting tangled in philosophical debates about the reality of particles that could in principle never be seen, he referred to the quarks as 'mathematical'. Some interpreted this to mean that Gell-Mann didn't think that the quarks were made of real 'stuff', entities that existed in reality and combined to give real effects.

Zweig was bolder (or, depending on your point of view, more reckless). In his second CERN preprint he had declared: 'There is also the outside chance that the model is a closer approximation to nature than we may think, and that fractionally charged aces abound within us.'[5]

---

Solid-state physicist Philip Anderson didn't believe Goldstone's theorem. It was transparently obvious from many practical examples in solid-state physics that Nambu–Goldstone bosons are not always produced when gauge symmetries are spontaneously broken. Symmetries were being broken all the time, yet solid-state physicists were hardly being overwhelmed by floods of massless, photon-like particles as a result. There were no massless particles generated inside superconductors, for example. Something was not quite right.

In 1963, Anderson suggested that the problems that the quantum field theorists were wrestling with could in some way resolve themselves:[6]

> It is likely, then, considering the superconducting analogue, that the way is now open . . . without any difficulties involving either zero-mass Yang–Mills gauge bosons or

> zero-mass [Nambu–]Goldstone bosons. These two types of bosons seem capable of 'canceling each other out' and leaving finite mass bosons only.

Could it really be that simple? Was this a case of two wrongs making a right? Anderson's paper provoked a minor controversy. As arguments and counter-arguments raged in the scientific press, a number of physicists took careful note.

There followed a series of papers detailing mechanisms for spontaneous symmetry-breaking in which the various massless bosons did indeed 'cancel each other out', leaving only massive particles. These were published independently by Belgian physicists Robert Brout and François Englert, English physicist Peter Higgs at Edinburgh University, and Gerald Guralnik, Carl Hagen, and Tom Kibble at Imperial College in London.* The mechanism is commonly referred to as the *Higgs mechanism* (or, in some quarters more concerned with the democracy of discovery, the Brout–Englert–Higgs–Hagen–Guralnik–Kibble – BEHHGK, or 'beck' mechanism).

The mechanism works like this. A massless field particle with spin 1 (a boson) moves at the speed of light and has two 'degrees of freedom', meaning that its wave amplitude can oscillate in two dimensions that are perpendicular (that is, transverse) to the direction in which it is travelling. If the particle is moving in the *z*-direction, say, then its wave amplitude can oscillate only in the *x*- and *y*-directions (left/right and up/down). For the photon, the two degrees of freedom are associated with left-

* These three papers were all published in the same volume (13) of the journal *Physical Review Letters* in 1964, on pp. 321–3, 508–9, and 585–7, respectively.

circular and right-circular polarization. These states can be combined to give the more familiar states of linear polarization: horizontal ($x$-direction) and vertical ($y$-direction). For light, there is no polarization in a third dimension.

To change this state it is necessary to introduce a background quantum field, often called the Higgs field, to break the symmetry.* The Higgs field is characterized by the shape of its *potential energy curve.*

The idea of a potential energy curve is relatively straightforward. Picture a pendulum swinging to and fro. As the pendulum rises in its swing it slows down, stops, and swings back the other way. At this point all the energy of its motion (its kinetic energy) has transferred into potential energy stored in the pendulum. As the pendulum swings back, the potential energy is released into the kinetic energy of motion and it picks up speed. At the bottom of its swing, with the pendulum pointing straight down, the kinetic energy is at a maximum and the potential energy is zero.

If we plot the value of the potential energy against the angle of displacement of the pendulum from the vertical, we get a parabola – see Figure 13(a). The minimum in this potential energy curve is clearly the point at which the displacement of the pendulum is zero.

The potential energy curve of the Higgs field is subtly different. Instead of the angle of displacement, we plot the displacement or value of the field itself. Towards the bottom

* Unlike other quantum fields we have so far encountered in this book, the Higgs is a 'scalar' field – it has magnitude at every point in space-time but no direction. In other words, it does not 'pull' or 'push' in any particular direction.

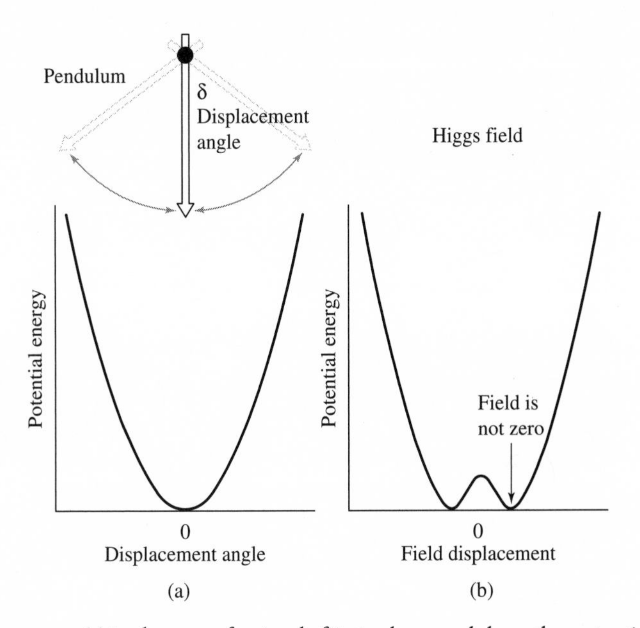


**FIGURE 13** (a) In the case of a simple frictionless pendulum, the potential energy curve is shaped like a parabola and zero potential energy corresponds to zero pendulum displacement. However, the potential energy curve for the Higgs field (b) is shaped differently. Now zero potential energy corresponds to a finite displacement (of the field itself), or what physicists call a non-zero vacuum expectation value.

of the curve there is a small bump, not unlike the top of a sombrero or a 'punt' (the indentation) in the bottom of a champagne bottle. The presence of this bump forces the symmetry to break. As the field cools and loses its potential energy, like the toppling pencil it randomly falls into a valley in the curve (the curve is actually three-dimensional). But this time the lowest point in the curve corresponds to a non-zero value of the field. Physicists refer to this as a non-zero vacuum expectation value. It represents a 'false' vacuum, meaning that the vacuum is not completely empty – it contains non-zero values of the Higgs field.

Breaking the symmetry creates a massless Nambu–Goldstone boson. This may now be 'absorbed' by the massless spin 1 field boson to create a third degree of freedom (forward/back). The wave amplitude of the field particle can now oscillate in all three dimensions, including the direction in which it is travelling. The particle acquires 'depth' (see Figure 14).

In the Higgs mechanism the act of gaining three-dimensionality is like applying a brake. The particle slows down to an extent which depends on the strength of its interaction with the Higgs field.

The photon does not interact with the Higgs field and continues to move unhindered, at the speed of light. It remains massless. Particles which do interact with the field acquire depth, gain energy, and slow down, the field dragging on them like molasses. The particle's interactions with the field are manifested as a resistance to the particle's acceleration.*

Does this sound vaguely familiar?

The inertial mass of an object is a measure of its resistance to acceleration. Our instinct is to equate inertial mass with the amount of substance that the object possesses. The more 'stuff' it contains, the harder it is to accelerate. The Higgs mechanism turns this logic on its head. *We now interpret the*

* Note that it is accelerated motion which is impeded. Particles moving at a constant velocity are not affected by the Higgs field. For this reason the Higgs field does not conflict with Einstein's special theory of relativity.

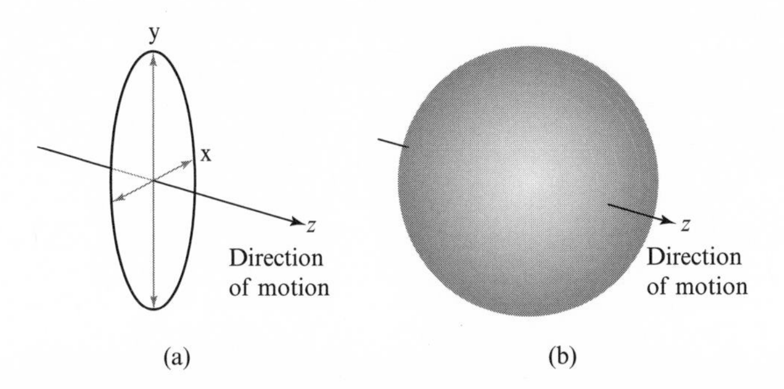


**FIGURE 14** (a) A massless boson moves at the speed of light and has just two transverse 'degrees of freedom', left/right (*x*) and up/down (*y*). On interacting with the Higgs field, the particle may absorb a massless Nambu–Goldstone boson, acquiring a third degree of freedom – forward back (*z*). Consequently, the particle gains 'depth' and slows down. This resistance to acceleration is the particle's mass.

*extent to which the particle's acceleration is resisted by the Higgs field as the particle's (inertial) mass.*

The concept of mass has vanished in a puff of logic. It has been replaced by interactions between otherwise massless particles and the Higgs field.

The Higgs mechanism did not win converts immediately. Higgs himself had some difficulties getting his paper published. He sent it initially to the European journal *Physics Letters* in July 1964, but it was rejected by the editor as unsuitable. Years later, Higgs wrote:[7]

> I was indignant. I believed that what I had shown could have important consequences in particle physics. Later, my colleague Squires, who spent the month of August 1964 at CERN, told me that the theorists there did not see the point

> of what I had done. In retrospect, this is not surprising: in 1964 . . . quantum field theory was out of fashion . . .

Higgs made some amendments to his paper and resubmitted it to the journal *Physical Review Letters*. It was sent to Nambu for peer review. Nambu asked Higgs to comment on the relationship between his paper and an article just published in the same journal (31 August 1964) by Brout and Englert. Higgs had not been aware of Brout and Englert's work on the same problem and acknowledged their paper in an added footnote. He also added a final paragraph to the main text in which he drew attention to the possibility of 'incomplete multiplets of scalar and vector bosons',[8] a rather obscure reference to the possibility of another, massive zero-spin boson, the quantum particle of the Higgs field.

This would come to be known as the Higgs boson.

Perhaps surprisingly, the Higgs mechanism had little immediate impact on those who might have benefited most from it.

---

Higgs was born in Newcastle upon Tyne, England, in 1929. In 1950 he graduated in physics at King's College, London and secured his PhD four years later. There followed spells at the University of Edinburgh and University and Imperial Colleges in London. He had returned to the University of Edinburgh in 1960 to take up a lectureship in Mathematical Physics. He married Jody Williamson, a fellow activist with the Campaign for Nuclear Disarmament, in 1963.

In August 1965 Higgs took Jody to Chapel Hill for a sabbatical period at the University of North Carolina. Their first son, Christopher, was born a few months later. Shortly afterwards Higgs received an invitation from Freeman Dyson to present a seminar on the Higgs mechanism at the Institute for Advanced Study. Higgs was wary of the likely reception of his theory at what had become commonly known as the Institutes' 'shotgun seminars', but when he delivered the seminar in March 1966 he emerged unscathed. Pauli had died in December 1958, but it is interesting to speculate if Higgs' arguments would have changed his attitude to Yang's unfortunate pleadings, a little over twelve years before.

Higgs took this opportunity to fulfil a long-standing request to give a seminar at Harvard University, and made his way there the next day. The audience was equally sceptical, with one Harvard theorist later admitting that they 'had been looking forward to tearing apart this idiot who thought he could get around the Goldstone theorem.'[9]

Glashow was in the audience, but it seems that he had by this time quite forgotten his earlier attempts to develop a unified electro-weak theory, a theory which predicted massless $W^+$, $W^-$, and $Z^0$ particles which needed somehow to be massive. 'His amnesia unfortunately persisted through 1966,' Higgs wrote.[10] In fairness to Glashow, Higgs was preoccupied with the application of his mechanism to the strong force.

But Glashow failed to put two and two together. It would be Glashow's former high school classmate Steven Weinberg (and, independently, Abdus Salam) who would eventually make the connection.

After receiving a bachelor's degree from Cornell University in 1954 Weinberg had begun his graduate studies at the Niels Bohr Institute in Copenhagen, returning to complete his PhD at Princeton University in 1957. He completed postdoctoral studies at Columbia University in New York and at the Lawrence Radiation Laboratory in California, before gaining a professorship at the University of California at Berkeley. He took a leave of absence to become a visiting lecturer at Harvard in 1966 and became a visiting professor at MIT the following year.

Weinberg had spent the previous couple of years working on the effects of spontaneous symmetry breaking in strong-force interactions described by an SU(2)×SU(2) field theory. As Nambu and Jona-Lasinio had found a few years before, the result of symmetry-breaking is that protons and neutrons acquire mass. Weinberg believed that the Nambu–Goldstone bosons so created could be approximated as the pions. At the time this all seemed to make sense and, far from trying to evade the Goldstone theorem, he now positively welcomed the predicted extra particles.

But now Weinberg realized that this approach wasn't going to bear fruit. It was at this point that he was struck by another idea:[11]

> At some point in the fall of 1967, I think while driving to my office at MIT, it occurred to me that I had been applying the right ideas to the wrong problem.

Weinberg had been applying the Higgs mechanism to the strong force. He now realized that the mathematical structures he had been trying to apply to strong-force interactions were precisely what were needed to resolve the problems with weak-force interactions and the massive bosons that these interactions implied. 'My God,' he exclaimed to himself, 'this is the answer to the weak interaction!'[12]

Weinberg was well aware that if the masses of the $W^+$, $W^-$, and $Z^0$ particles were added by hand, as in Glashow's SU(2)×U(1) electro-weak field theory, then the result was rendered unrenormalizable. He now wondered if breaking the symmetry using the Higgs mechanism would endow the particles with mass, eliminate the unwanted Nambu–Goldstone bosons, and yield a theory that could in principle be renormalized.

There remained the problem of the weak neutral currents, interactions involving the neutral $Z^0$ particle for which there was still no experimental support. He decided to avoid this problem altogether by restricting his theory to leptons – electrons, muons, and neutrinos. He had by now become wary of the hadrons, particles affected by the strong force, and especially the strange particles, the principal ground for the experimental exploration of weak-force interactions.

Neutral currents would still be predicted, but in a model consisting only of leptons these currents would involve the neutrino. The neutrino had proved difficult enough to find experimentally in the first place, and Weinberg may have figured that finding weak-force neutral currents involving these particles would present such insurmountable experimental challenges that he could predict them with little fear of contradiction.

Weinberg published a paper detailing an electro-weak unified theory for leptons in November 1967. This was an

SU(2)×U(1) field theory reduced to the U(1) symmetry of ordinary electromagnetism by spontaneous symmetry-breaking, giving mass to the $W^+$, $W^-$, and $Z^0$ particles whilst leaving the photon massless. He estimated the mass-scales of the weak-force bosons, about 85 times the proton mass for the W particles and 96 times the proton mass for the $Z^0$. He was not able to prove that the theory was renormalizable but felt confident that it was.

In 1964, Higgs had referred to the possibility of the existence of a Higgs boson, but this was not in relation to any specific force or theory. In his electro-weak theory, Weinberg had found it necessary to introduce a Higgs field with four components. Three of these would give mass to the $W^+$, $W^-$, and $Z^0$ particles. The fourth would appear as a physical particle – the Higgs boson. What had earlier been a mathematical possibility had now become a prediction. Weinberg even estimated the strength of the coupling between the Higgs boson and the electron. The Higgs took a critically important step towards becoming a 'real' particle.

In Britain, Abdus Salam had been introduced to the Higgs mechanism by Tom Kibble. He had worked earlier on an SU(2)×U(1) electro-weak field theory and immediately saw the possibilities afforded by spontaneous symmetry-breaking. When he saw a preprint of Weinberg's paper applying the theory to leptons he discovered that both he and Weinberg had independently arrived at precisely the same model. He decided against publishing his own work until he had had an opportunity properly to incorporate hadrons. But, try as he might, he could not get around the problem of the weak neutral currents.

Both Weinberg and Salam believed that the theory was renormalizable, but neither was able to prove this. They were also unable to predict the mass of the Higgs boson.

---

Nobody took much notice. Those few who did pay attention tended to be critical. The mass problem had been fixed through some 'smoke-and-mirrors' trick involving a hypothetical field which implied another hypothetical boson. It seemed as though the quantum field theorists were continuing to play games with fields and particles, according to obscure rules that few understood.

Particle physicists simply ignored them and got on with their science.

# 5

# I Can Do That

*In which Gerard 't Hooft proves that Yang–Mills field theories can be renormalized and Murray Gell-Mann and Harald Fritzsch develop a theory of the strong force based on quark colour*

Aside from the absurd fractional electric charges, there was another big problem with the quark model. As constituents of 'matter particles' such as protons and neutrons, the quarks were required to be fermions, with half-integral spins. This meant that, according to Pauli's exclusion principle, the hadrons could not accommodate more than one quark in each of the possible quantum states.

But the quark model insisted that the proton should consist of two up-quarks and a down-quark. This was a bit like saying that an atomic orbital should contain two spin-up electrons and one spin-down electron. It was just not possible. The symmetry properties of the electron wavefunctions forbade it. There could be only two electrons, one spin-up and one spin-down. There was no room for a third. Likewise, if the quarks were fermions, then there could be no room for two up-quarks in the proton.

This problem had been identified shortly after publication of Gell-Mann's first quark paper. Physicist Oscar Greenberg suggested in 1964 that quarks might actually be *parafermions*, which was tantamount to saying that the quarks could be distinguished by other 'degrees of freedom' apart from the one for which the quantum numbers were up, down, and strange. As a result there would be different kinds of up-quarks, for example. So long as they were of different kinds, two up-quarks could sit happily alongside each other in a proton without occupying the same quantum state.

But there were problems with this model, too. Greenberg's solution opened the door for baryons to behave like bosons, condensing into a single macroscopic quantum state like a beam of laser light. This was just not acceptable.

Yoichiro Nambu toyed with a similar scheme, suggesting that there could be first two, then three different kinds of up-, down-, and strange-quarks. A young graduate student from Syracuse University in New York, Korean-born Moo-Young Han, wrote to him in 1965 elaborating this idea. Together they wrote a paper which was published later that year.

This was not a simple extension of Gell-Mann's quark theory, however. Han and Nambu introduced a new type of 'quark charge' that was different from electric charge. The two up-quarks in the proton were now distinguished by their different quark charges, thereby avoiding conflict with Pauli's exclusion principle. They reasoned that the force holding the quarks together inside the larger nucleons is based on a local SU(3) symmetry, not to be confused with the global SU(3) symmetry that lies behind the Eightfold Way.

They also decided to use this opportunity to rid the quark theory of its fractional electric charges, introducing instead overlapping SU(3) triplets with electric charges of +1, 0, and −1, alongside the quark charge.

Nobody took much notice. Han and Nambu had taken a big step towards the ultimate solution, but the world was not yet ready.

---

Glashow finally returned to the problems of his SU(2)×U(1) electro-weak field theory in 1970, in the company of two Harvard postdoctoral associates, Greek physicist John Iliopoulos and Italian Luciano Maiani. Glashow had first met Iliopoulos at CERN and had been impressed with his efforts to find ways to renormalize a field theory of the weak force. Maiani arrived at Harvard with some curious ideas about the strength of weak-force interactions. All three realized that their interests converged.

At this stage none was aware of Weinberg's 1967 paper applying spontaneous symmetry-breaking and the Higgs mechanism in an electro-weak theory of leptons.

Glashow, Iliopoulos, and Maiani now wrestled with the theory once again. Adding the masses of the $W^+$, $W^-$, and $Z^0$ particles by hand produced unruly divergences in the equations that rendered the theory unrenormalizable. Then there was the problem of the weak neutral currents. For example, the theory predicted that a neutral kaon should decay by emission of a $Z^0$ boson, changing the strangeness of the particle in the process and producing two muons – a weak neutral current. However, there was absolutely no experimental evidence for this decay mode. Rather than abandon the $Z^0$ altogether, the physicists tried to work out why this particular mode might be suppressed.

The muon neutrino had been discovered in 1962, adding a fourth lepton alongside the electron, electron neutrino, and muon. The physicists started to tinker with the model of four leptons and three quarks, initially adding more leptons. But Glashow had actually published a paper in 1964 speculating on the possible existence of a fourth quark, which he had called the *charm-quark*. This seemed to make more sense. Nature would surely demand a balance between the numbers of leptons and the numbers of quarks. A model of four leptons and four quarks had a much more pleasing symmetry.

The theorists added a fourth quark to the mix, a heavy version of the up-quark with charge $+\frac{2}{3}$. They realized that by doing this, they had switched off the weak neutral currents.

The weak neutral currents could arise through decays involving the $Z^0$ and more complex decays involving emission of both $W^-$ and $W^+$ particles. In both cases the end-result is the same – two oppositely charged muons, $\mu^-$ and $\mu^+$. This latter decay path is depicted in Figure 15(a). Here, a neutral kaon (shown as a combination of down- and anti-strange-quarks) emits a virtual $W^-$ particle, and the down-quark (charge $-\frac{1}{3}$) transforms into an up-quark (charge $+\frac{2}{3}$). The virtual $W^-$ particle decays into a muon and muon anti-neutrino.

The resulting up-quark can then be thought to emit a virtual $W^+$ particle, turning into a strange-quark in the process. The $W^+$ particle decays into a positive muon and muon

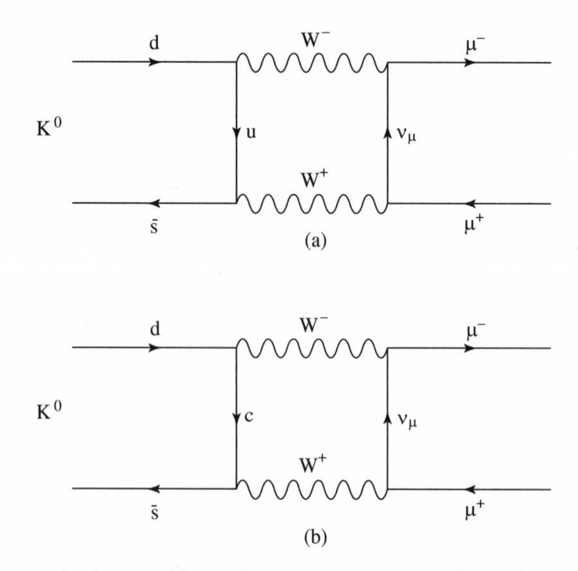


**FIGURE 15** (a) A neutral kaon decays into two muons through a complex mechanism involving emission of $W^-$ and $W^+$ particles. There is no net change in charge and so this is a weak neutral current. (b) The decay path defined in (a) is cancelled out by this alternative decay path involving the charm-quark (denoted here by *c*).

neutrino. This is referred to as a 'one loop contribution' to the net result, which involves decay of a neutral kaon into positive and negative muons.

There was, in principle, no reason why this example of a neutral current shouldn't be observed. However, the common decay modes of neutral kaons produce pions, not muons. Somehow the decay path to muons was being suppressed. Glashow, Iliopoulos, and Maiani realized that an entirely analogous decay path involving the charm-quark would do the trick – Figure 15(b). A difference in signs relating to these two possible decay paths meant that they virtually cancel each other out. Caught like a rabbit in headlights, the neutral kaon can't decide which way to jump, until it's too late.

It was a neat solution. Kaons, the principal ground for experimental studies of weak-force interactions which should have exhibited weak neutral currents, almost never did so because of alternative decay modes involving the charm-quark.

Excited by their discovery, the physicists piled into a car and headed across to MIT and the office of American physicist Francis Low, who had also been working on the problem. Weinberg joined them and together they debated the merits of this new Glashow–Iliopoulos–Maiani (GIM) mechanism.

What followed was an extraordinary failure of communication.

Almost all the ingredients for a unified theory of the weak and electromagnetic forces were assembled in the minds of the theorists gathered in Low's office. Weinberg had figured out how to apply spontaneous symmetry-breaking using the Higgs mechanism in an SU(2)×U(1) field theory of leptons, allowing the masses of the field particles to be calculated rather than input by hand. Glashow, Iliopoulos, and Maiani had found a potential solution to the problem of weak neutral currents in strange-particle decays and held out the promise that the SU(2)×U(1) theory could be extended to weak-force interactions involving the hadrons. But they were still inputting the masses of the field particles by hand and struggling with divergences.

Glashow, Iliopoulos, and Maiani knew nothing of Weinberg's 1967 paper, and Weinberg said nothing about it. He later confessed to having a 'psychological barrier' against his earlier work, specifically in relation to the problem of demonstrating that the electro-weak theory could be renormalized.[1] He also did not look kindly on the charm-quark proposal. What Glashow, Iliopoulos, and Maiani were invoking was not just one new particle, part of an extended family of particles of possibly dubious relevance, but an entirely new collection of 'charmed' baryons and mesons. If the charm-quark existed, then the Eightfold Way would simply be a subset of a much larger representation with many charmed members.

It was an awful lot to swallow just to explain the *absence* of weak neutral currents in strange-particle decays. 'Of course, not everyone believed in the predicted existence of charmed hadrons,' said Glashow.[2]

---

There could be no further progress until someone was able to show that the Weinberg–Salam electro-weak theory could be renormalized.

Dutch theorist Martinus Veltman had studied mathematics and physics at the State University of Utrecht, becoming professor at the University in 1966. He began to work on the problems of renormalizing Yang–Mills field theories in 1968.

High-energy physics was not a popular subject for research in the Netherlands. This led to a certain sense of isolation. But this suited Veltman's purposes, as it meant that he did not have to defend his choice of unfashionable research topics.

At the beginning of 1969 a young student, Gerard 't Hooft, was assigned to him to complete a pre-doctoral thesis (known colloquially as a 'scriptie'). Veltman shied away from getting his young student to work on Yang–Mills theories as he judged this subject too risky, and unlikely to lead to gainful employment. But after successfully completing his pre-doctoral thesis, 't Hooft was offered a position at the University so that he could study for a PhD. 't Hooft expressed a desire to continue working with Veltman.

Veltman still judged Yang–Mills field theories to be filled with dangers. He had made some considerable progress with renormalization, but the problem was extremely stubborn. But 't Hooft felt strongly that this would prove to be fertile ground for his doctoral thesis. Veltman initially suggested an alternative topic, but 't Hooft would not be diverted.

They were an unlikely pairing. Veltman's was a larger-than-life, no-nonsense personality, proud of his achievements although indifferent to the physics community's general lack of interest. 't Hooft was slightly built and rather self-effacing, his modesty masking a mind of singular sharpness.

In his 1997 book *In Search of the Ultimate Building Blocks*, 't Hooft would introduce Veltman with an anecdote. One day Veltman entered an elevator that was already full. As the button was pressed, the elevator alarm warned that it was overloaded. All eyes turned to Veltman, relatively large of girth and one of the last to enter. But while others might have mumbled an embarrassed apology and stepped out, Veltman was having none of it. He understood Einstein's equivalence principle, which underpins the general theory of relativity – if a person falls freely he will not feel his own weight. He knew what he needed to do.

'When I say "yes", then press!' he exclaimed.[3]

He then proceeded to jump into the air. 'Yes!' he yelled.

Someone pressed the button, and the elevator began its ascent. By the time Veltman had returned to the floor, the elevator had picked up enough speed to continue its journey. 't Hooft was in the elevator.

Sometime in the autumn or winter of 1970–71, Veltman and 't Hooft were walking between buildings on the University campus.

'I do not care what and how,' Veltman told his student, 'But what we must have is at least one renormalisable theory with massive charged vector bosons, and whether that looks like nature is of no concern, [those] are details that will be fixed later by some model freak. In any case, all possible models have been published already.'[4]

'I can do that,' 't Hooft said, quietly.

Knowing full well the stubbornness of the problem, and that others – such as Richard Feynman – had tried, and failed, 't Hooft's declaration greatly surprised him. He almost walked into a tree.

'What do you say?' he asked.

'I can do that,' 't Hooft repeated.

Veltman had been working on this problem for so long that he simply could not believe its solution was as easy as 't Hooft was making out. He was understandably sceptical.

'Write it down and we will see,' he said.

But 't Hooft had learned about spontaneous symmetry-breaking at a summer school in Cargèse, Corsica, in 1970. By late 1970 he had shown in his first paper that Yang–Mills field theories containing massless particles could be renormalized. 't Hooft was confident that applying spontaneous symmetry-breaking would enable Yang–Mills theories with massive particles to be renormalized, too.

Within a short time he had indeed written it down.

Veltman was unhappy with 't Hooft's use of the Higgs mechanism. He was particularly concerned that the presence of a background Higgs field, pervading the entire universe, should reveal itself through gravitational effects.*

And so they argued back and forth. In the end, 't Hooft decided to give his thesis advisor the results of his theoretical manipulations without showing explicitly where they had come from. Veltman knew well enough, but was content just to check the veracity of 't Hooft's results.

Some years earlier Veltman had developed a novel approach to performing complex algebraic manipulations using a computer program he called Schoonschip, 'clean ship' in Dutch.† This was one of the first computer algebra systems, able to manipulate mathematical equations in symbolic form. He now took 't Hooft's results with him to Geneva to check them on the CERN computer.

Veltman was excited but remained sceptical. Looking over the results as he set up his computer program, he decided to

* By contributing specifically to a cosmological constant, first introduced as a 'fudge factor' by Einstein in his gravitational field equations. In the lambda-CDM model of Big Bang cosmology, the cosmological constant (lambda) controls the rate of expansion of space-time.

† This is a Dutch naval expression meaning to clean up a messy situation. Veltman later claimed he chose this name to annoy everybody not Dutch.

drop some factors of four that had appeared in 't Hooft's equations, factors that could be traced to the Higgs boson. He thought the factors of four were just crazy. He set up his program, and ran it without these factors.

He was soon calling 't Hooft on the phone, declaring: 'It nearly works. You just have some factors of two wrong.'[5]

't Hooft wasn't wrong. 'So then he realized that even the factor of four was right,' 't Hooft explained, 'and that everything canceled in a beautiful way. By that time he was as excited about it as I had been.'

't Hooft had quite independently (and by sheer coincidence) re-created the broken SU(2)×U(1) field theory that Weinberg had developed in 1967, and had now shown how it could be renormalized. 't Hooft had thought to apply the field theory to the strong force, but when Veltman asked a CERN colleague if he knew of any other applications of an SU(2)×U(1) theory, he was pointed in the direction of Weinberg's paper. Veltman and 't Hooft now realized that they had developed a fully renormalizable quantum field theory of electro-weak interactions.

It was a major breakthrough. '. . . the psychological effect of a complete proof of renormalizabiity has been immense,' wrote Veltman some years later.[6] In fact, what 't Hooft had done was demonstrate that Yang–Mills gauge theories in general are renormalizable. Local gauge theories are actually the only class of field theories that can be renormalized.

't Hooft was just 25 years old. Initially, Glashow didn't understand the proof. Of 't Hooft he said: 'Either this guy's a total idiot or he's the biggest genius to hit physics in years.'[7] Weinberg didn't believe it, but when he saw that a fellow theorist was taking it seriously he decided to look more closely at 't Hooft's work. He was quickly convinced.

't Hooft was appointed as an assistant professor at Utrecht.

Now *all* the ingredients were available. A renormalizable, spontaneously broken SU(2)×U(1) field theory of the weak and electromagnetic forces was now at hand. The masses of the W and $Z^0$ bosons emerged 'naturally' from the application of the Higgs mechanism. Some anomalies remained, but 't Hooft had pointed out in a footnote to his paper that these did not render the theory unrenormalizable. 'Of course,' he wrote years later, 'this should be interpreted as saying that renormalizability can be restored by adding an appropriate amount of various kinds of fermions (quarks), but I admit that I also thought that perhaps this was not even necessary.'[8]

The anomalies that remained could be removed by adding more quarks to the model.

---

What hopes now for a field theory of the strong force?

Gell-Mann collected the 1969 Nobel Prize for physics for his many contributions, most notably the discoveries of strangeness and the Eightfold Way. His achievements were listed in the ceremonial presentation speech, delivered by Ivar Waller, a member of the Nobel Committee for Physics. Waller also mentioned the quarks, explaining that, though eagerly sought, they had not been found. He graciously conceded that the quarks were nonetheless of great 'heuristic' value.

Gell-Mann had now to come to terms with the celebrity status bestowed upon Nobel Laureates. Inundated with requests to attend meetings and to submit papers, a writing process he had always found difficult now became impossible. He even missed the deadline for submission of his own Nobel Prize lecture for publication in the Swedish Academy's *Le Prix Nobel*.* It was one among many missed deadlines.

In the summer of 1970 he retreated to Aspen, Colorado, with his family. But this was a retreat from commitments, not physics. They vacationed with the families of other physicists at the grounds of the Aspen Center for Physics.

The Center was tailor-made for Nobel Laureates seeking freedom from distraction. It had been established in 1962 by the Aspen Institute for Humanistic Studies following an approach from two physicists. Their idea was to create a facility that would offer a peaceful, relaxed, unstructured environment where physicists could escape the administrative demands of their day-to-day academic jobs and simply talk physics among themselves. The Institute set aside part of its Aspen Meadows campus, amidst a grove of aspen trees on the edge of the town.

It was in Aspen that Gell-Mann encountered Harald Fritzsch, a fervent believer in the quark model, astonished to discover that Gell-Mann was curiously ambivalent about his own 'mathematical' creation.

Fritzsch had been born in Zwickau, south of Leipzig in East Germany. Together with a colleague he had defected from Communist East Germany, escaping from the authorities in

* The Nobelprize.org website states flatly that: 'Professor Gell-Mann has presented his Nobel Lecture [on 11 December 1969], but did not submit a manuscript for inclusion in this volume.'

Bulgaria in a kayak fitted with an outboard motor. They had travelled 200 miles down the Black Sea to Turkey.

He had begun studying for a doctorate in theoretical physics at the Max Planck Institute for Physics and Astrophysics in Munich, West Germany, where one of his professors was Heisenberg. In the summer of 1970 he passed through Aspen on his way to California.

As a student in East Germany, Fritzsch had become convinced that quarks must lie at the heart of a quantum field theory of the strong nuclear force. These things were much more than mathematical devices. They were real.

Gell-Mann was impressed by the young German's enthusiasm and agreed to have Fritzsch join him at Caltech, visiting about once a month. Together they began to work on a field theory constructed from quarks. When Fritzsch completed his graduate studies in West Germany in early 1971, he transferred to Caltech.

Fritzsch had triggered a small earthquake, shaking the foundations of Gell-Mann's conservative attitude to the quarks. This was more than just a psychological earthquake: Fritzsch's arrival at Caltech on 9 February 1971 coincided with a real earthquake that struck the San Fernando Valley early that morning near Sylmar, with a magnitude of 6.6 on the Richter scale. 'In memory of that occasion,' Gell-Mann later wrote, 'I left the pictures on the wall askew, until they were further disturbed by the 1987 earthquake.'[9]

Gell-Mann organized grants for himself and for Fritzsch and in the autumn of 1971 they both travelled to CERN. It was here that William Bardeen, son of John Bardeen of the BCS theory of superconductivity, told them about some anomalies in the calculated decay rates of neutral pions. Bardeen had spent some time at Princeton working with Stephen Adler on this calculation. They had showed that the model of fractionally charged quarks predicted a decay rate which came out a factor of three too low compared with the measured value. Adler had gone on to show that the Han–Nambu model of integral-charged quarks actually did a better job of predicting the measured rate.

Gell-Mann, Fritzsch, and Bardeen now worked together to explore the options. They wanted to see if it was possible to reconcile the results on neutral pion decay with a variation of the original model of fractionally charged quarks.

As Han and Nambu had suggested, what they needed was a new quantum number. Gell-Mann decided to call this new quantum number 'colour'. In this new scheme, quarks would possess three possible colour quantum numbers: blue, red, and green.*

Baryons would be constituted from three quarks of different colour, such that their total 'colour charge' is zero and their product is 'white'. For example, a proton could be thought to consist of a blue up-quark, a red up-quark and a green down-quark ($u_b u_r d_g$). A neutron would consist of a blue up-quark, a red down-quark and a green down-quark ($u_b d_r d_g$). The mesons, such as pions and kaons, could be thought to consist of

* In their original scheme Gell-Mann, Fritzsch, and Bardeen called them red, white, and blue (inspired by the French national flag). However, it soon became clear that the three primary colours would work better as, when blended, they produce the colour white. To avoid confusion I have adopted the currently accepted terminology from the outset.

coloured quarks and their anti-coloured anti-quarks, such that the total colour charge is zero and the particles are also 'white'.

It was a neat solution. The different quark colours provided the extra degree of freedom and meant that there was no violation of the Pauli exclusion principle. Tripling the number of different types of quarks meant that the decay rate of the neutral pion was now accurately predicted. And nobody could expect to see the colour charge revealed in experiments as this was a property of quarks and the quarks are 'confined' inside white-coloured hadrons. Colour could not be seen because nature demands that all observable particles are white.

'We gradually saw that that [colour] variable was going to do everything for us!' Gell-Mann explained. 'It fixed the statistics, and it could do that without involving us in crazy new particles. Then we realized that it could also fix the dynamics, because we could build an SU(3) gauge theory, a Yang–Mills theory, on it.'[10]

By September 1972, Gell-Mann and Fritzsch had elaborated a model consisting of three fractionally charged quarks which could take three 'flavours' – up, down, and strange – and three colours, bound together by a system of eight coloured gluons, the carriers of the strong 'colour force'. Gell-Mann presented the model at a conference on high-energy physics held to mark the opening of the National Accelerator Laboratory in Chicago.

But he was already beginning to have second thoughts. Once more troubled particularly by the status of the quarks and the mechanism by which they are permanently confined, Gell-Mann gave the theory a somewhat muted fanfare. He mentioned a variation of the model featuring a single gluon. He emphasized that the quarks and gluons were 'fictitious'.

By the time he and Fritzsch came to write up the lecture, they had been overtaken by their doubts. 'In preparing the written version,' he later wrote, 'unfortunately, we were troubled by the doubts just mentioned, and we retreated into technical matters.'[11]

This failure of courage is not so difficult to understand. If the coloured quarks really were permanently confined inside 'white' baryons and mesons, such that their fractional electric charges and their colour charges can never be seen, then it could be argued that all speculation about their properties was inherently idle.

The theorists were now very close to a grand synthesis: a combination of quantum field theories based on an SU(3)×SU(2)×U(1) symmetry – what would become known as the Standard Model. This would be a synthesis which would set the theoretical stage for experimental particle physics for the next thirty years. This hesitancy was simply a deep breath before the plunge.

In fact, tantalizing evidence for the existence of quarks had emerged just a few years earlier from high-energy collisions involving electrons and protons. The results of experiments conducted at the Stanford Linear Accelerator Center (SLAC) in California hinted strongly that the proton consists of point-like constituents.

But it was not clear that these point-like constituents were quarks. Even more puzzling, the results also suggested that, far from being held in a tight grip inside the proton, the constituents behaved as though they were entirely free to roam around inside their larger hosts. How was this meant to be compatible with the idea of quark confinement?

The theorists' work was almost complete. The Standard Model was almost in place. It was now the turn of the experimentalists.

# PART II
# Discovery

# 6
# Alternating Neutral Currents

*In which protons and neutrons are shown to have an internal structure and the predicted neutral currents of the weak nuclear force are found, and then lost, and then found again*

Cosmic rays produce some of the highest-energy particle collisions ever observed, much higher in some instances than can be achieved even with today's particle colliders.* But the origin of the rays is mysterious, and the particles and energies involved in triggering events are unknown. Successful cosmic ray experiments rely on chance detection of new particles or new processes, detection that can prove difficult to replicate.

Despite the success of cosmic ray experiments in uncovering the positron, the muon, pions, and kaons in the two decades spanning the 1930s to the early 1950s, further progress

* The energies of cosmic ray particles are typically between 10 MeV and 10 GeV, but very occasionally particles of incredibly high energies are recorded. On 15 October 1991 a cosmic ray particle was detected in Utah with an energy of approximately 300 million TeV. This was referred to as the 'Oh-My-God' particle, thought to be a proton accelerated to speeds very close to that of light.

in particle physics had to await the development of ever-more powerful man-made particle accelerators.

The first accelerators were built in the late 1920s. These were linear accelerators, producing acceleration of electrons or protons by passing them through a linear sequence of oscillating electric fields. One such accelerator was used in 1932 by John Cockcroft and Ernest Walton to produce high-speed protons which were then fired at stationary targets, transmuting the target nuclei in the first artificially induced nuclear reactions.*

American physicist Ernest Lawrence invented an alternative accelerator design in 1929. This involved using a magnet to confine a stream of protons to move in a spiral whilst accelerating them to higher and higher speeds using an alternating electric field. He called it the *cyclotron*.

Lawrence was also something of a showman, with grand ambitions. There followed a succession of larger and larger machines, culminating in 1939 with a design for a gargantuan super-cyclotron with a magnet weighing two thousand tons. Lawrence estimated that this would deliver proton energies of 100 million electron volts (100 MeV), on the threshold of the energies required for protons to penetrate the nucleus. Lawrence approached the Rockefeller Foundation with requests for support. His pitch was greatly strengthened when, in the middle of a game of tennis, he was informed that he had just won the 1939 Nobel Prize for physics.

* These were rather inaccurately reported as experiments which 'split the atom'.

With the outbreak of war, Lawrence's cyclotron technology was diverted to the problem of separating quantities of uranium-235 sufficient to produce the atom bomb that was dropped on Hiroshima. The electromagnetic isotope separation facility Y-12, constructed at Oak Ridge in eastern Tennessee, was based on Lawrence's cyclotron design.*

The magnets used at Y-12 were 250 feet long and weighed between three thousand and ten thousand tons. Their construction exhausted America's supply of copper, and the US Treasury had to loan the Manhattan project fifteen thousand tons of silver to complete the windings. The magnets required as much power as a large city and were so strong that workers could feel the pull of magnetic force on the nails in their shoes. Women straying close to the magnets would occasionally lose their hairpins. Pipes were pulled from the walls. Thirteen thousand people were employed to run the plant, which began operation in November 1943.

This was the first example of what would become known as 'big science'.

The cyclotron used a constant magnetic field strength and fixed-frequency electric field and so had an inherent limit in terms of particle energies of about 1000 MeV (or 1 giga electron volt, GeV). To access yet higher energies, it is necessary to drive the accelerated particles in bunches around a circular track along which both the magnetic and electric fields are synchronously varied. Early examples of such

* Electromagnetic separation was not the only technique used. A huge gaseous diffusion plant (K-25) and a thermal diffusion plant were also constructed at Oak Ridge.

*synchrotrons* included the Bevatron, a 6.3 GeV accelerator built in 1950 at the Radiation Laboratory in Berkeley, California, and the Cosmotron, a 3.3 GeV machine built in 1953 at Brookhaven National Laboratory in New York.

Other countries began to get in on the act. On 29 September 1954, eleven western European countries ratified a convention to establish the *Conseil Européen pour la Recherche Nucléaire* (the European Council for Nuclear Research, or CERN).* Three years later a 10 GeV proton synchrotron was inaugurated by the Soviet Union's Joint Institute for Nuclear Research in Dubna, 120 kilometres north of Moscow. CERN soon followed in 1959 with a 26 GeV proton synchrotron in Geneva.

Funding for high-energy physics in America was greatly increased as the race for Cold War technological supremacy reached white heat in the 1960s. The Alternating Gradient Synchrotron was constructed at Brookhaven in 1960, capable of operating at 33 GeV. It seemed evident that the future development of particle physics lay in the hands of synchrotron designers, pushing the technology to ever greater collision energies.

So, when construction of a new $114-million 20 GeV *linear* electron accelerator commenced in 1962 at Stanford University in California, many particle physicists dismissed it as an irrelevant machine, capable only of second-rate experiments.

* This was renamed the *Organisation Européenne pour la Recherche Nucléaire* (European Organization for Nuclear Research) when the provisional Council was dissolved. However, the acronym OERN was judged to be clumsier than CERN, so the original acronym was retained.

But some physicists recognized that the emphasis on ever higher-energy hadron collisions had come at the cost of subtlety. The synchrotrons were used to accelerate protons and smash them into stationary targets, including other protons. As Richard Feynman explained, proton–proton collisions were '. . . like smashing two pocket watches together to see how they are put together.'[1]

The Stanford Linear Accelerator Center (SLAC) was built on 400 acres of Stanford University grounds about 60 kilometres south of San Francisco. It reached its 20 GeV design beam energy for the first time in 1967. The three-kilometre accelerator is linear, rather than circular, because bending electron beams into a circle using intense magnetic fields results in dramatic energy loss through emission of X-ray synchrotron radiation.

When an electron collides with a proton, three different types of interaction may result. The electron may bounce relatively harmlessly off the proton, exchanging a virtual photon, changing the electron's velocity and direction but leaving the particles intact. This, so-called 'elastic' scattering yields electrons with relatively high scattered energies clustered around a peak.

In a second type of interaction, the collision with the electron may exchange a virtual photon which kicks the proton into one or more excited energy states. The scattered electron comes away with less energy as a result, and a chart of scattered energy vs. yield shows a series of peaks or 'resonances' corresponding to different excited states of the proton. Such scattering is 'inelastic', as new particles (such as pions) may be created, although both electron and proton emerge intact from the interaction. In essence, the energy of the collision, and of the exchanged virtual photon, has gone into the creation of new particles.

The third type of interaction is called 'deep inelastic' scattering, in which much of the energy of the electron and the exchanged virtual photon goes into destroying the proton completely. A spray of different hadrons results and the scattered electron recoils, now with considerably less energy.

Studies of deep inelastic scattering at relatively small angles from a liquid hydrogen target began at SLAC in September 1967. They were carried out by a small experimental group including MIT physicists Jerome Friedman and Henry Kendall and Canadian-born SLAC physicist Richard Taylor.

They focused their attention on the behaviour of something called the 'structure function' as a function of the difference between the initial electron energy and the scattered electron energy. This difference is related to the energy lost by the electron in the collision or the energy of the virtual photon that is exchanged. They saw that as the virtual photon energy was increased, the structure function showed marked peaks corresponding to the expected proton resonances. However, as the energy increased further, these peaks gave way to a broad, featureless plateau that fell away gradually as it extended well into the range of deep inelastic collisions.

Rather curiously, the shape of the function appeared to be largely independent of the initial electron energy. The experimentalists didn't understand why.

But American theorist James Bjorken did. Bjorken had obtained his doctorate at Stanford University in 1959 and had recently returned to California after a spell at the Niels

Bohr Institute in Copenhagen. Just before SLAC was completed, he had developed an approach to predicting the outcomes of electron–proton collisions using a rather esoteric approach based on quantum field theory.

In this model, it was possible to think of the proton in two distinct ways. It could be considered as a solid 'ball' of material substance, with mass and charge distributed evenly. Or it could be thought of as a region of largely empty space containing discrete, point-like charged constituents, much as the atom had been shown in 1911 to be empty space containing a tiny, positively charged nucleus.

These two very different ways of thinking about the structure of the proton would produce very different scattering results. Bjorken had understood that electrons of sufficient energy could penetrate the interior of a 'composite' proton and collide with its point-like constituents. In the region of deep inelastic collisions, the electrons would be scattered in greater numbers and at larger angles, and the structure function would behave in the way now being revealed by the experiments.

Bjorken had drawn back from declaring that such point-like constituents might be quarks. The quark model was still treated with derision by the physics community and there were alternative theories available that were better regarded. Arguments over the interpretation of the data broke out even within the group of MIT-SLAC physicists. Consequently, the physicists did not rush to declare the results as evidence for the existence of quarks.

And there the matter rested for another ten months.

Richard Feynman visited SLAC in August 1968. After working on the weak nuclear force and aspects of quantum gravity, he had decided to turn his attention back to high-energy physics. His sister Joan lived in a house near the SLAC facility, and during visits to her he would take the opportunity to 'snoop around' at SLAC to find out what was happening in the field.

He heard about the work of the MIT-SLAC group on deep inelastic scattering. A second round of experiments was about to begin, but the physicists were still puzzling over the interpretation of the data from the previous year.

Bjorken was out of town, but his new postdoctoral research associate Emmanuel Paschos told Feynman about the behaviour of the structure function and asked him what he thought. When Feynman saw the data he declared: 'All my life I've looked for an experiment like this, one that can test a field theory of the strong force!'[2] He figured it out that night in his motel room.

He believed that the behaviour the MIT-SLAC physicists had seen was related to the distribution of momentum of point-like constituents deep inside the proton. Feynman called these constituents 'partons' – literally 'parts of protons' – to avoid getting entangled with any specific model for the interior of the proton.*

'I've really got something to show youse guys,' Feynman declared to Friedman and Kendall the next morning. 'I figgered

* Gell-Mann was unimpressed. He called them 'put-ons'. In truth, 'partons' are not just quarks – they can be both quarks and the gluons that transmit the colour force between them.

it all out in my motel room last night!'[3] Bjorken had already arrived at most of the conclusions that Feynman now drew, and Feynman acknowledged his priority. But, once again, Feynman was describing the physics in a far simpler, yet richer, more visual way. When he returned to SLAC in October 1968 to deliver a lecture on the parton model, it was like setting a fire. Nothing breeds confidence in a bold idea than its enthusiastic advocacy by a Nobel Laureate.

Were partons actually quarks? Feynman didn't know and didn't care, but Bjorken and Paschos soon had a detailed model of partons based on a triplet of quarks.

Further studies of deep inelastic scattering of electrons from neutrons at SLAC and results from studies of the scattering of neutrinos from protons at CERN provided further supporting evidence. By mid-1973, quarks had officially 'arrived'. They might have been conceived partly in jest as a strange quirk of nature, but they had now taken a decisive step towards acceptance as real constituents of the hadrons.

Some important questions remained unanswered. The behaviour of the structure function could only be properly understood if the quarks were assumed to be individually bouncing around inside the proton or neutron, completely independently of each other. And yet, the 20-GeV electrons had struck the individual quarks, resulting in the destruction of the target nucleon hosts, so how come no free quarks had been liberated?

It didn't make sense. If the strong force kept the quarks so tightly bound inside the nucleons that they were forever 'confined' and could never be seen, how could it be that inside the nucleons they were moving about apparently so freely?

---

By the end of 1971, a fully fledged quantum field theory of electro-weak interactions had been worked out, and the theorists' confidence was growing. Symmetry-breaking using the Higgs mechanism could explain the distinction between electromagnetism and the weak nuclear force, which would otherwise be the same universal electro-weak force. Symmetry-breaking had left the photon massless whilst giving mass to the carriers of the weak force. The weak force demanded two charged force carriers, the $W^+$ and $W^-$ particles, and a neutral force carrier, the $Z^0$. If the $Z^0$ existed then interactions involving its exchange could be expected to be manifested in the form of weak neutral currents.

If the theory was correct, then neutral kaons could be expected to exhibit weak neutral currents which also involved a change in strangeness. The rather embarrassing absence of such strangeness-changing currents was now explained by invoking the GIM mechanism and the existence of a fourth, charm-quark.

The theorists turned their attentions to other sources of weak neutral currents which did not involve a change in strangeness number and began to urge the experimentalists to search for them. The best candidate events appeared to involve interactions between muon neutrinos and nucleons: protons and neutrons. In the collision of a muon neutrino and a neutron, for example, the exchange of a virtual $W^-$ particle turns the muon neutrino into a negative muon and the neutron into a proton. This is a charged current. Exchange of a virtual $Z^0$ particle leaves both the muon neutrino and the neutron intact – a neutral current (see Figure 16). If both processes occur then evidence for weak neutral currents could be obtained by scattering muon neutrinos from nucleons and looking for events in which no muon is produced. Weinberg estimated that for every 100 charged current events there should be somewhere between 14 and 33 neutral current events.

The problem was that neutrinos are extremely light, neutral particles which leave no trace in particle detectors. Such detectors depend on the passage of *charged* particles dislodging

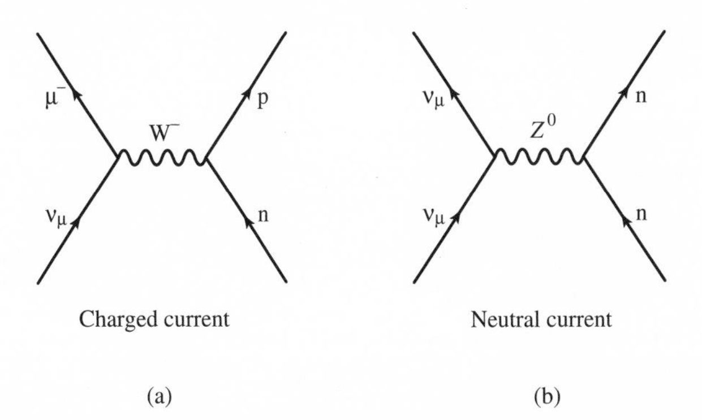


**FIGURE 16** (a) A neutron collides with a muon neutrino and a virtual $W^-$ particle is exchanged. This turns the neutron into a proton and the neutrino into a muon. This is a charged 'current'. However, the same collision may also involve exchange of a virtual $Z^0$ particle, (b). No particle changes identity and no muon is produced. This 'muonless' event is a neutral current.

electrons from the atoms of the detector material, leaving a tell-tale trail of charged ions in their wake. The first detector of this type was invented by Scottish physicist Charles Wilson in 1911. In Wilson's 'cloud chamber' the particle tracks are made visible through the condensation of water vapour around the ions that are left behind.

The cloud chamber was superseded in the early 1950s by the bubble chamber, invented by American physicist Donald Glaser, but the principles are very similar. A bubble chamber is filled with liquid held close to its boiling point. A charged particle passing through the liquid again leaves a trail of ions and electrons in its wake. If the pressure above the liquid is then lowered, the liquid begins to boil. But it will boil first along the trail of ions left behind, forming a series of bubbles which make the track visible. The tracks can then be photographed and the pressure increased to stop any further boiling of the liquid.

The advantage of the bubble chamber is that the chamber liquid can also serve as the target for particles from an accelerator. Most bubble chambers used liquid hydrogen, but heavier liquids such as propane and freons (the liquids used in older refrigerators) could also be used.

The only signature of a 'muonless' event of the kind Weinberg was seeking would be a spray of hadrons suddenly appearing in the detector, seemingly out of nowhere. But there were many other, rather mundane, explanations for such mysterious sprays of hadrons. Muon neutrinos might strike atoms in the detector walls, chipping off stray neutrons which could go on to produce random hadrons in the detector. Events occurring 'upstream' of the detector could produce neutrons which then produce hadrons. And if a muon produced in a charged current event was scattered with a large recoil angle, there was a good chance that it would be missed altogether. Background events such as these could easily be miscounted as genuine muonless events, and therefore erroneously identified as weak neutral currents.

The experimentalists were extremely wary of the difficulties involved in any such search. A list of experimental priorities drawn up by CERN physicists in November 1968 put the W particles at the top, but the search for weak neutral currents was a humble eighth. 'The fact is that, up until 1973, there was no firm evidence in favour of neutral currents and plenty of evidence against them,' wrote Oxford physicist Donald Perkins.[4]

However, by the spring of 1972 the enormous theoretical advances that had been achieved pushed the search to the top of the agenda. The physicists began to think that it might be possible to provide a definitive answer.

A large and growing international collaboration led by CERN physicist Paul Musset, Andre Lagarrigue from the accelerator laboratory in Orsay, and Donald Perkins utilized 'Gargamelle', the world's largest heavy liquid bubble chamber. Gargamelle had been funded by the French atomic energy commission, built in France and installed at CERN in 1970 alongside the 26 GeV proton synchrotron.* It had taken six

* It was named for the mother of the giant Gargantua, from French Renaissance author Francois Rabelais' sixteenth-century novels *The Life of Gargantua* and *Pantagruel*.

years to construct, and was designed specifically to study collisions involving neutrinos.

Gargamelle had been in operation for almost a year, and had thrown up lots of muonless events that had been dismissed as background 'noise' produced by stray neutrons. The experimentalists now began to look at these events with renewed interest.

The challenge was to distinguish genuine muonless events produced by weak neutral currents from those produced by background neutrons, large-angle muon scattering, and misidentification. It was a painstaking and rather thankless task, but by late 1972 many physicists in the Gargamelle collaboration, which by now included physicists from seven European laboratories and guests from America, Japan, and Russia, were beginning to believe that they had found something. But opinion within the collaboration was divided, not so much about the reality or otherwise of the neutral currents themselves, but rather about whether the evidence they had gathered was sufficiently compelling.

In the meantime, a second search had begun in America. The world's largest proton synchrotron had been constructed at the National Accelerator Laboratory (NAL)* in Chicago, reaching its design energy of 200 GeV in March 1972. Italian physicist Carlo Rubbia at Harvard, Alfred Mann at the University of Pennsylvania, and David Cline at the University of Wisconsin now used beams of muon neutrinos generated by the synchrotron to look for muonless events. The CERN team

* This was renamed the Fermi National Accelerator Laboratory (Fermilab) in 1974.

was ahead, but their preliminary reports were inconclusive. Rubbia was ambitious and determined to get there first.

Finding muonless events was easy. Proving that they derived from weak neutral currents was hard. When Musset presented further preliminary data in early 1973 there was no fanfare, no claim to have made the discovery that they were all pursuing.

The NAL team's advantage allowed them an opportunity to catch up. Their synchrotron was more powerful, able to create more muon neutrino scattering events in a shorter time. Their detector also provided a larger target mass than Gargamelle, improving the chances of detecting the scattering events. These factors helped to reduce the impact of background neutrons, but there was nothing that could be done about muons scattered at large angles 'escaping' without being detected. Rubbia and his team at Harvard sought to account for this contribution using a computer simulation, subtracting a theoretical estimate of the contribution from the number of muonless events measured experimentally, and so arriving at the number of genuine muonless events.

It was a clumsy compromise, and Mann and Cline were deeply suspicious. Rubbia, aware that the CERN physicists were building their body of evidence, was in a hurry.* Mann and Cline understood only too well how such pressures could

* The CERN physicists had also by this time found a single 'gold-plated' weak neutral current event among some of the older photographs from Gargamelle. This involved the interaction of a muon anti-neutrino with an electron, a much rarer process but one that is free from background contamination. It was unambiguous evidence, but it was still only one photograph. Eventually, after searching through almost one and a half million photographs, only *three* such events were found.

easily lead the physicists to self-delusion, to convince themselves of the existence of something that actually wasn't there. They urged caution.

News of the NAL physicists' progress reached CERN in July 1973. Rubbia wrote to Lagarrigue claiming that they had accumulated 'approximately one hundred unambiguous [neutral current] events'.[5] He went on to suggest that the groups publish their findings simultaneously. Lagarrigue politely declined. The CERN physicists had identified genuine muonless events in the collisions of muon neutrinos with nucleons and had estimated the ratio of neutral to charged current events to be 0.21. For collisions involving muon anti-neutrinos the ratio was 0.45. The physicists now moved to declare that weak neutral currents had at last been found, and submitted a paper to the journal *Physics Letters*. It was published in September.

The NAL group had found the combined ratio of neutral to charged currents for both muon neutrino and anti-neutrino collisions to be 0.29, in good agreement with the CERN results.*

At this critical juncture Rubbia's visa expired and, despite holding a professorship at Harvard, he was threatened with deportation. During his appeal hearing at the offices of the Immigation and Naturalization Service in Boston he lost his temper. He was on a plane out of the country within 24 hours.

With Rubbia out of the picture, the NAL collaborators began to back-track. Their paper, which they had submitted

* The ratio of muon neutrinos and anti-neutrinos in the NAL experiments was of the order of two to one. The weighted average of the CERN ratios for muon neutrinos and anti-neutrinos is therefore 0.29.

to the journal *Physical Review Letters* in August, was rejected by peer reviewers concerned that the problems of eliminating erroneous muonless events had not been properly addressed. Cline and Mann now rebuilt their detector, intending to settle the matter, one way or the other.

The genuine muonless events promptly disappeared, with ratios of neutral to charged current events falling as low as 0.05. The NAL physicists became convinced that they had been misled by their earlier results.

Rubbia was also a prominent figure at CERN and decided to stir up trouble. He advised CERN Director-General Willibald Jentschke that the Gargamelle collaboration had made a big mistake. CERN was still very much in the shadow of its more prestigious American rivals and its international reputation had suffered some setbacks from previous errors. Many European physicists were inclined to think that the Gargamelle result must be wrong and one senior CERN physicist staked half the contents of his wine cellar on a bet against the result. Appalled at the thought that CERN's reputation was about to suffer another blow, Jentschke summoned the Gargamelle physicists to a meeting. It seemed like an inquisition.

But, although the Gargamelle physicists were shaken by these developments, they were resolute. They chose to stand by their conclusions. Encountering Jentschke in the lift at CERN, Perkins offered reassurance. 'I knew the group had gone through the event analysis many times and for almost a year we had searched intensively for some other explanation for the effects observed, without success,' Perkins explained. 'So I thought the result was absolutely solid, and [Jentschke] should just ignore rumours from across the Atlantic. I don't know if my words reassured him, but he got out of the lift with a smile on his face.'[6]

Rubbia returned to NAL in early November and together the NAL physicists began to draft a rather different paper, declaring that weak neutral currents had not been found in contradiction with the recent report from CERN and the predictions of the electro-weak theory.

There following a rather embarrassing *volte face*. By mid-December 1973 the NAL physicists had realized that pions creeping in from other neutrino collisions had been misidentified in their detector as muons. This effect had virtually eliminated the count of muonless events. Weak neutral currents were back. Cline now had to admit 'the distinct possibility that a muonless signal of order ten per cent is showing up in the data'.[7] He could not find a way to make these events disappear. The NAL team decided to resubmit their original paper, with suitable modification. It was published in *Physical Review Letters* in April 1974.

Some in the physics community referred jokingly to the discovery of 'alternating neutral currents'.

By mid-1974, other laboratories had confirmed the result and the confusion had cleared. Weak neutral currents were an established experimental fact.

But the implications of this discovery were of even greater significance. Weak neutral currents implied the existence of 'heavy photons' responsible for carrying the weak force. And if no neutral currents could be found in strange-particle decays, this must be because they are suppressed by the GIM mechanism.

In other words, there must also be a fourth quark.

# 7
# They Must Be Ws

---

*In which quantum chromodynamics is formulated, the charm-quark is discovered, and the W and Z particles are found, precisely where they were predicted to be*

The pieces of the jigsaw were now falling into place. The puzzle of the existence of point-particles moving freely about inside nucleons, revealed in the experiments on deep inelastic scattering at SLAC, was shown not to be a puzzle at all. It was a direct consequence of the nature of the strong nuclear force, which behaves rather counter-intuitively.

When imagining the nature of an interaction governed by a force between two particles, we tend to think of examples such as gravity or electromagnetism, in which the force grows stronger as the particles get closer together.* But the strong nuclear force doesn't behave in this way. The force exhibits what is known as *asymptotic freedom*. In the asymptotic limit of

* Think about your childhood experience of pushing the north poles of two bar magnets together. The resistance you feel increases as you push the magnets closer together.

zero separation between two quarks the particles feel no force and are completely 'free'. As the separation between them increases beyond the boundary of the nucleon, however, the strong force tightens its grip and holds them in check.

It is as if the quarks were fastened to the ends of strong elastic. When the quarks are close together inside a nucleon, the elastic is relaxed and there is little or no force between them. The force is experienced only when we try to pull the quarks apart and so stretch the elastic (see Figure 17).

In late 1972 Princeton theorist David Gross had set out to show that asymptotic freedom was simply impossible in a quantum field theory. With the help of his student Frank Wilczek he managed instead to prove precisely the opposite. Quantum field theories based on local gauge symmetries *can*

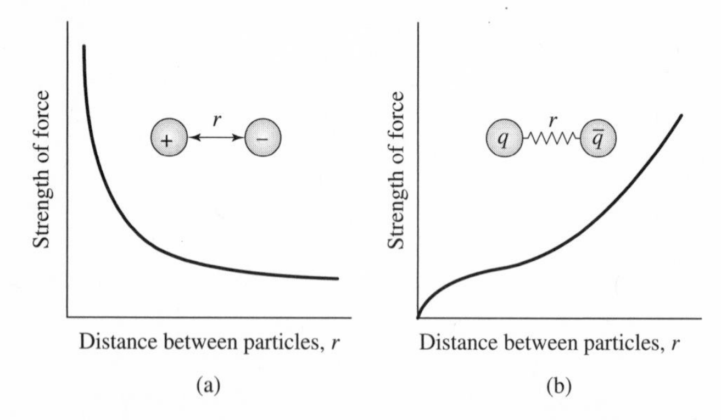


**FIGURE 17** (a) The electromagnetic force of attraction between two electrically charged particles increases as the particles move closer together. But the colour force that binds quarks inside hadrons behaves rather differently, (b). In the limit of zero separation between a quark and an anti-quark (for example), the force falls to zero. The force increases as the quarks are separated.

accommodate asymptotic freedom. A young Harvard graduate student called David Politzer independently made the same discovery. Their papers were published back-to-back in the June 1973 issue of *Physical Review Letters*.*

Gell-Mann retreated once again to the Aspen Center that June, clutching preprints of the Gross–Wilczek and Politzer papers. He was joined by Fritzsch and Heinrich Leutwyler, a Swiss theorist from the University of Bern on study leave at Caltech. Together they developed a Yang–Mills quantum field theory of three coloured quarks and eight coloured, massless gluons.† To account for asymptotic freedom, the gluons were now *required* to carry colour charge. No tricks involving a Higgs-like mechanism were required.

The new theory needed a name. In 1973 Gell-Mann and Fritzsch had been referring to it as quantum hadron dynamics, but the following summer Gell-Mann thought he had come up with a better name. 'The theory had many virtues and no known vices,' he explained. 'It was during a subsequent summer at Aspen that I invented the name quantum chromodynamics, or QCD, for the theory and urged it upon Heinz Pagels and others.'[1]

* In fact, 't Hooft had already concluded that Yang–Mills gauge theories could show this counter-intuitive behaviour, but he was busy working on renormalization at this time and did not follow it up.

† Massless gluons? What about the claims of Heisenberg and Yukawa, that the carriers of the strong force should be large, massive particles? This would indeed be a requirement if the strong force were like gravity or electromagnetism, but it's not. The asymptotically free colour force can be quite happily carried by massless particles. Like the quarks, these are confined inside hadrons, which is why they are not as ubiquitous as photons.

A great synthesis, combining the theories of the strong and electro-weak forces in a single SU(3)×SU(2)×U(1) structure, seemed at last to be at hand.

But whilst asymptotic freedom could explain why quarks interact only very weakly inside hadrons, it does not explain why the quarks are confined. Various picturesque models were devised. In one of these, the gluon fields surrounding the quarks are imagined to form narrow tubes or 'strings' of colour charge between the quarks as they separate. As the quarks are pulled apart, the string first tenses and then stretches, the resistance to further stretching increasing with increasing separation.

Eventually the string breaks, but at energies sufficient to conjure quark–anti-quark pairs spontaneously from the vacuum. So, pulling a quark from the interior of a nucleon, for example, cannot be done without creating an anti-quark which will immediately pair with it to form a meson, and another quark which will take its place inside the nucleon. The end result is that energy is channelled into the spontaneous creation of a meson, and no individual quarks can be observed. Quarks are not so much confined as never, but never, seen without a chaperone.*

The cost, in energy terms, of an isolated or 'naked' colour charge is vast. In principle, the energy of a single, isolated quark is infinite. The quark rapidly accumulates a covering of virtual gluons in an attempt to mask the colour charge, and

* Analogies of this kind are colourful (no pun intended) but remain speculative. To this day, confinement remains a problem in QCD that has yet to be resolved.

the energy increases. It costs a lot less energy to mask the charge either by pairing with an anti-quark of the same colour or combining with two other quarks of different colour such that the net colour charge is zero and the resulting host particle is 'white'.

However, the quark charge cannot be masked completely. To do this we would need somehow to pile the quarks right on top of each other. But quarks are just like electrons – they are quantum particles with wave as well as particle properties. According to Heisenberg's uncertainty principle, pinning the positions of the quarks down in this manner would lead to an infinite uncertainty in their momenta. This implies the possibility of infinite momentum, which is just as costly.

Nature settles for a compromise. The colour charge cannot be completely masked but the energy, manifested in the associated gluon fields, can be reduced to manageable proportions. This energy is nevertheless substantial. It turns out that the (hypothetical) masses of the up- and down-quarks are quite small, between 1.5 and 3.3 MeV and between 3.5 and 6.0 MeV, respectively.* The measured mass of a proton is 938 MeV, the neutron mass is about 940 MeV. The combined mass of two up-quarks and a down-quark would be something like 4.5–9.9 MeV. So where does the rest of the proton mass come from? It comes from the *energy* of the gluon fields inside the proton.

'Does the inertia of a body depend on its energy content?' Einstein had asked in 1905. The answer is yes. About 99 per cent of the mass of protons and neutrons is energy carried by

* These quark mass data are taken from C. Amsler *et al.*, *Physics Letters B*, **667** (2008), p. 1.

the massless gluons that hold the quarks together. 'Mass, a seemingly irreducible property of matter, and a byword for its resistance to change and sluggishness,' wrote Wilczek, 'turns out to reflect a harmonious interplay of symmetry, uncertainty, and energy.'[2]

---

Glashow visited Brookhaven in August 1974, once more to urge the experimentalists to search for the charm-quark. American physicist Samuel Ting was listening. He was preparing to use the 30 GeV Alternating Gradient Synchrotron (AGS) to study high-energy proton–proton collisions and watch carefully for electron–positron pairs emerging amidst the chaos of hadrons produced.

When the data revealed that electron–positron pairs were piling up in a narrow 'resonance' at an energy of around 3 GeV, the experimentalists were not sure what to make of this. They sought to eliminate obvious sources of error and re-checked their analysis. It made no difference. The peak remained stubbornly fixed at 3.1 GeV, and stubbornly narrow. They began to suspect that this might be new physics.

Ting was cautious. He had acquired a reputation for showing up errors in other physicists' experiments, and he didn't want to fall victim to the same treatment. He resisted the pressure to publish the results until they had had a chance to reconfirm the data.

Meanwhile, over on the West Coast, Stanford University physicist Roy Schwitters had a problem. The Stanford Positron Electron Asymmetric Rings (SPEAR), used to collide accelerated electrons and positrons, had come on stream at SLAC in mid-1973. Schwitters had found an error in one of the computer programs used to analyse data from the SPEAR experiments. When he corrected it, data re-analysed from experiments carried out in June 1974 now showed hints of structure – small bumps at 3.1 and 4.2 GeV. The project leader, American physicist Burton Richter, was eventually persuaded to reconfigure SPEAR for collision energies around 3.1 GeV, so that they could go back and have another look.

By November 1974, it was clear that both Ting's group at Brookhaven and Richter's group at SLAC had discovered the same new particle, a meson formed from a charm-quark and an anti-charm-quark. Ting's group had resolved to call it the J-particle, Richter's group called it the ψ (psi). This joint discovery was subsequently referred to as the 'November revolution'.

There followed something of a hiatus over priority. Neither group would concede priority by adopting the name given by the other, and the particle is today still called the J/ψ. Ting and Richter shared the 1976 Nobel Prize for physics.

---

The discovery of the J/ψ was a triumph for theoretical and experimental physics. It also served to neaten the structure of the fundamental particles – the foundation of what was fast becoming the 'Standard Model' of particle physics.

There were now two 'generations' of fundamental particles, each consisting of two leptons and two quarks and the particles responsible for carrying forces between them. The electron, electron neutrino, up-quark, and down-quark form the first generation. The muon, muon neutrino, strange-quark, and charm-quark form the second generation, differentiated from the first by their masses. The photon carries the electromagnetic force, the W and Z particles carry the weak nuclear force, and eight coloured gluons carry the strong nuclear force or colour force between the coloured quarks.

But by the spring of 1977, overwhelming evidence had accumulated for an even heavier version of the electron – called the tau lepton. It was not what physicists really wanted to hear.

A tau lepton demanded a tau neutrino and, inevitably, speculation mounted that there are actually *three* generations of leptons and quarks. American physicist Leon Lederman found the upsilon (*Υ*) at Fermilab in August 1977. This is a meson consisting of what had by then come to be known as a bottom quark and its anti-quark. With a mass of about 4.2 GeV, the bottom quark is a heavier, third-generation version of the down- and strange-quarks with a charge of $-\frac{1}{3}$. It was assumed that the final member of the third generation – the top quark – was heavier still and would be found as soon as colliders capable of the requisite collision energies could be built.

Although it had made something of a surprise appearance, the third generation of leptons and quarks was readily absorbed into the Standard Model (see Figure 18). At a symposium organized at Fermilab in August 1979, evidence was presented for the appearance of quark and gluon 'jets' produced in electron–positron annihilation experiments. These are directed sprays of hadrons produced from the formation of a quark–anti-quark pair in which an energetic gluon is also

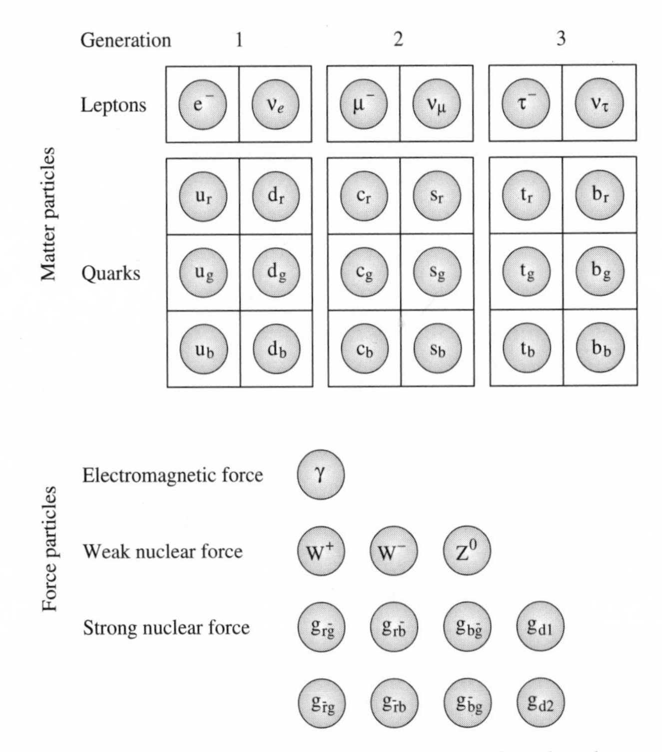


**FIGURE 18** The Standard Model of particle physics describes the interactions of three generations of matter particles through three kinds of force, mediated by a collection of field particles or 'force carriers'.

'liberated' from one of the quarks. Such tell-tale 'three-jet events' provide the most striking evidence yet found for both quarks and gluons.

The top quark was still missing, as was direct evidence for the W and Z particles, the carriers of the weak force. As the Standard Model became the new orthodoxy, Glashow,

Weinberg, and Salam learned that they had been awarded the 1979 Nobel Prize for physics for their work on electro-weak unification.

The race was now on to find the remaining particles needed to complete the set. In his Nobel Prize lecture, Weinberg explained that the electro-weak theory predicted masses for the W and Z particles of about 83 GeV and 94 GeV, respectively.*

Back in June 1976, CERN had commissioned its Super Proton Synchrotron (SPS), a 6.9 kilometre circumference proton accelerator capable of generating particle energies up to 400 GeV. A month before its commissioning, these particle energies had already been surpassed by the proton accelerator at Fermilab, which had reached 500 GeV. But smashing particles into stationary targets results in a substantial waste, as energy is carried away by recoiling particles. In this kind of arrangement, the energy that can usefully be channelled into the creation of new particles increases only as the square-root of the particle energy in the beam.

This meant that collisions involving particles accelerated even to the energies now available from the SPS or Fermilab accelerator could be expected to produce new particles only of much lower energy. To reach the energies predicted for the W and Z particles would require an accelerator considerably larger than any yet built.

There was an alternative. The idea of colliding two beams of accelerated particles had been developed in the 1950s. If the

* If the mass of the proton is taken to be 938 MeV, then these correspond to about 88 and 100 times the proton mass, respectively.

accelerated particles were passed into two linked storage rings, in beams travelling in opposite directions, then the beams could be brought into head-on collision. Now *all* the energy of the accelerated particles could be channelled into the creation of new particles.

Such particle *colliders* were first constructed in the 1970s. SPEAR was an early example, but it utilized head-on collisions between leptons (electrons and positrons). In 1971 CERN completed construction of the Intersecting Storage Rings (ISR), a hadron (proton–proton) collider which used the 26 GeV proton synchrotron as the source of accelerated protons. The protons would first be accelerated in the synchrotron before being passed into the ISR, where they would be brought into collision. However, the peak collision energy – 52 GeV – was still insufficient to reach the W and Z particles.

In April 1976 a study group was assembled at CERN to report on the next major construction project, called the Large Electron–Positron (LEP) collider. This was to be built in a 27-kilometre circular tunnel passing beneath the Swiss–French border near Geneva. It would use the SPS to accelerate electrons and positrons to speeds close to that of light before injecting them into the collider ring. Collisions would involve particles (in this case electrons) and their anti-particles (positrons), which would be circulated in opposite directions in a single ring. The initial design energy was 45 GeV for each particle beam which, when combined, would produce head-on collision energies of 90 GeV, bringing the LEP just within reach of the W and Z particles.

American physicist Robert Wilson, the director of Fermilab, had an even grander vision. He wanted to build a hadron collider capable of reaching collision energies of 1 TeV (1000 GeV, a terra electron volt or a trillion electron volts). It would eventually become known as the 'Tevatron'. Such a collider would require as yet untried and untested superconducting magnets to accelerate the particles. And it was no more than a proposal.

Such was the situation faced by high-energy particle physicists in 1976. CERN's SPS could accelerate particles to 400 GeV and its ISR could reach collision energies of 52 GeV, neither of which was sufficient to find the W and Z particles. The LEP would in principle be capable of finding them but this machine would not be available until 1989. Fermilab's Main Ring could accelerate particles to 500 GeV, still insufficient to find the W and Z particles. The Tevatron, capable in theory of reaching collision energies of 1 TeV, was on the drawing board.

The physicists didn't have the patience to wait. 'The pressure to discover the W and Z was so strong,' recalled CERN physicist Pierre Darriulat, 'that the long design, development and construction time of the LEP project left most of us, even the most patient among us, unsatisfied. A quick (and hopefully not dirty) look at the new bosons would have been highly welcome.'[3] Patience was also wearing thin among the physicists at Fermilab.

What the physicists on both sides of the Atlantic needed to do was figure out how they could stretch their existing facilities to the all-important energy regime.

---

One possible solution had emerged in the late 1960s. It was possible – in principle – to turn an accelerator into a hadron collider by creating beams of protons and anti-protons that circulate around an accelerator ring in opposite directions. The beams could then be brought into head-on collision. A proton–proton collider required two intersecting rings, with protons in each ring travelling in opposite directions, but proton–anti-proton collisions could be engineered in a single ring. And it would be possible to achieve collision energies equal to twice the highest accelerator energies.

But this was no straightforward matter. Anti-protons are produced by colliding high-energy protons with stationary targets (such as copper). A million such collisions are required to produce a single anti-proton. Worse still, the anti-protons are produced with a broad range of energies, too broad to be accommodated in a storage ring. Only a small fraction of the anti-protons so produced would 'fit' in the ring, greatly reducing both the intensity of the anti-proton beam and the beam *luminosity*, a measure of the number of collisions that the beam can produce.

To make a beam of anti-protons sufficiently luminous for successful proton–anti-proton collider experiments would require that the anti-proton energies be somehow 'gathered' and concentrated around the desired beam energy.

Fortunately, Dutch physicist Simon van der Meer had figured out how to do precisely this. Van der Meer had graduated in engineering from the Delft University of Technology in 1952. He had worked for the Philips electronics company in the Netherlands for a few years before joining CERN in 1956. At CERN he became an accelerator theorist, primarily concerned with the practical application of theoretical principles to the design and operation of particle accelerators and colliders.

Van der Meer had performed some speculative experiments using the ISR in 1968 but did not publish an internal report on his findings until four years later. The reason for his tardiness was simple: the physics he was pursuing seemed vaguely mad. In his report he wrote: 'The idea seemed too far-fetched at the time to justify publication.'[4]

His 1968 experiments had hinted that it might indeed be possible to concentrate anti-protons with an initial spread of energies to the much narrower energy range needed to fit in a storage ring. The technique involved using 'pick-up' electrodes to sense anti-protons with energies that strayed from the desired beam energy and sending a signal to a 'kicker' electrode on the other side of the ring to nudge these particles back into line. The signals passed from pick-up to kicker electrodes are like a shepherd's whistled instructions to a sheep dog. On receiving the instructions, the dog barks the stray sheep back into line, and allows the flock to be neatly escorted into the pen.

Van der Meer called the technique 'stochastic cooling'. The word stochastic simply means 'random', and the cooling refers not to the temperature of the beam but to the random motions and the energy spread of the particles contained within it. By repeating the process many millions of times, the beam would gradually converge on the desired beam energy. In 1974 van der Meer carried out some further tests of stochastic cooling using the ISR. The results were not substantial, but they were sufficient to suggest that the principle worked.

In the meantime, Carlo Rubbia had set aside his disappointment at having been beaten to the discovery of weak neutral currents by CERN physicists. Rubbia had secured his PhD at the Scuola Normale in Pisa, Italy, in 1959. He had worked on aspects of muon physics at Columbia University before joining CERN in 1961. In 1970 he was appointed to a professorship at Harvard, spending one academic term a year there and the balance of his time back at CERN. His globetrotting had attracted the award of a nickname by his Harvard students, who called him the 'Alitalia professor'.

Rubbia was also stubborn, single-minded, ambitious, and notoriously difficult to work with.* He had resolved that he would *not* be beaten to the discovery of the W and Z particles.

Together with colleagues from Harvard, in mid-1976 Rubbia had submitted proposals to Wilson to convert Fermilab's 500 GeV proton synchrotron into a proton–anti-proton collider. Wilson had declined, preferring to focus his energies instead on garnering support for the Tevatron. The stochastic cooling technique seemed like a long-shot. If it didn't work, potentially valuable time on the synchrotron would be lost. Wilson agreed to a half-million-dollar experiment with a small-scale machine to discover if the technique would work.

Rubbia simply took his proposal back with him to CERN, where it got a much more positive reception from Leon van Hove, then CERN Director-General. By June 1978 further CERN trials of stochastic cooling had yielded results that

* Martinus Veltman wrote, of Rubbia: 'When he was Director of CERN, he changed secretaries at the rate of one every three weeks. This is less than the average survival time of a sailor on a submarine or destroyer in World War II . . .' See Veltman, p. 74.

were greatly encouraging, and van Hove was ready to take a gamble. This was an opportunity for CERN to discover new particles, an achievement that had for some years been the preserve of American facilities. Besides, had van Hove not agreed, Rubbia would most probably have gone back to Leon Lederman, who had taken over at Fermilab following Wilson's resignation in February.* 'Most likely, if CERN hadn't bought Carlo [Rubbia]'s idea, he would have sold it to Fermilab,' Darriulat explained.[5]

Rubbia was given the go-ahead to form a collaborative team of physicists to design the elaborate detector facility that would be required to discover the W and Z particles. As this was to be constructed in a large excavated area on the SPS the collaboration was called Underground Area 1, or UA1. The team would grow to include some 130 physicists.

Six months after the decision a second, independent collaboration, UA2, was formed under Darriulat's leadership. This would be a smaller collaboration, consisting of some 50 physicists, designed to provide friendly competition with UA1. The UA2 detector facility would be less elaborate (it would not be able to detect muons, for example), but would nevertheless be able to provide independent corroboration of the UA1 findings.

* Wilson had run into problems with the funding for Fermilab and had quit in frustration. As it turned out, after an exhaustive review of the options in November 1978, Lederman decided that the risks associated with using the existing facility as a proton–anti-proton collider were too great. He was not prepared to gamble in the way that van Hove was, and decided to throw his weight behind a renewed effort to secure funding for the Tevatron.

Proton and anti-proton beam energies of 270 GeV would combine in the SPS to produce collisions with a total energy of 540 GeV, well in excess of the energies required to reveal the W and Z particles.

---

After some delays, UA1 and UA2 finally began logging data in October 1982. It was anticipated that collisions producing the W and Z particles would be very rare, so both detector facilities were set up so that they would respond only to selected collisions meeting pre-programmed criteria. The collider would produce several thousand collisions per second over a period of two months. Only a handful of W- and Z-producing events were expected.

The detector facilities were programmed to identify events involving the ejection of high-energy electrons or positrons at large angles to the beam direction. Electrons carrying energies up to about half the mass of the W would be the signature of the decay of $W^-$ particles. High-energy positrons would likewise signal the decay of $W^+$ particles. Measured energy imbalances (differences between the energies of the particles going into the collision versus those coming out) would signal the concomitant production of anti-neutrinos and neutrinos, which could not be detected directly.

Preliminary results were presented at a workshop in Rome in early January 1983. Rubbia, uncharacteristically nervous, made the announcement. From the several thousand million collisions that had been observed, UA1 had identified six events that were candidates for W-particle decays. UA2 had identified four candidates. Though somewhat tentative, Rubbia was convinced: 'They look like Ws, they smell like Ws, they must be Ws.'[6] 'His talk was spectacular,' wrote Lederman. 'He had all the goods and the showmanship to display them with a passionate logic.'[7]

On 20 January 1983, CERN physicists packed into the auditorium to hear two seminars delivered by Rubbia for UA1 and Luigi Di Lella for UA2. A press conference was called on 25 January. The UA2 collaboration preferred to reserve judgement, but judgement was soon forthcoming. The W particles had been found, with energies close to the predicted 80 GeV.

The UA1 discovery of the $Z^0$, with a mass around 95 GeV, was announced on 1 June 1983. This was based on the observation of five events – four producing electron–positron pairs and one producing a muon pair. The UA2 collaboration had accumulated a few candidate events by this time but preferred to wait for results from a further experimental run before going public. UA2 eventually reported eight events producing electron–positron pairs.

By the end of 1983, UA1 and UA2 between them had logged about a hundred $W^{\pm}$ events and a dozen $Z^0$ events, revealing masses around 81 GeV and 93 GeV, respectively.

Rubbia and van der Meer shared the 1984 Nobel Prize for physics.

---

It had been a long journey, one that was, arguably, begun with Yang and Mills' seminal 1954 work on an SU(2) quantum field theory of the strong force. This was the theory which predicted the massless bosons that had so irked Pauli. In 1957 Schwinger had speculated that the weak nuclear force is mediated by three field particles, and his student Glashow had subsequently reached for an SU(2) Yang–Mills field theory to accommodate them.

The discovery of the Higgs mechanism in 1964 had shown how massless bosons of this kind could acquire mass. Weinberg and Salam had gone on to apply the Higgs mechanism to electro-weak symmetry-breaking in 1967–68. The resulting theory had been shown to be renormalizable in 1971. And now the carriers of the weak force had been found, precisely where they had been expected.

The very existence of the W and Z particles with the predicted masses provided rather compelling evidence that the SU(2)×U(1) electro-weak theory was basically right. And if the theory was right, then interactions with an all-pervasive energy field (the Higgs field) were responsible for endowing the weak-force carriers with mass. And if the Higgs field exists, so too must the Higgs boson.

But finding the Higgs boson was going to require an even bigger collider.

# 8

# Throw Deep

*In which Ronald Reagan throws his weight behind the Superconducting Supercollider, but when the project is cancelled by Congress six years later all that remains is a hole in Texas*

What the physicists had learned from their experience of electro-weak unification could be applied again to a larger problem. The electro-weak theory implied that at some time shortly after the big bang, the temperature of the universe would have been so high that the weak nuclear force and the electromagnetic force would have been indistinguishable. There was instead a single electro-weak force carried by massless bosons.

This is known as the 'electro-weak epoch'. As the universe cooled, the background Higgs field 'crystallized' and the higher gauge symmetry of the electro-weak force was broken (or, more correctly, 'hidden'). The massless bosons of electromagnetism (photons) continued unimpeded, but the weak-force bosons interacted with the Higgs field and gained mass to become the W and Z particles. The upshot is that in terms of interaction strengths and scales, the weak and electromagnetic forces now look very different.

In 1974 Weinberg, American theorist Howard Georgi, and Australian-born physicist Helen Quinn showed that the strengths of the interactions of all three particle forces become near-equal at energies between a hundred billion and a hundred million billion GeV.* These energies, corresponding to temperatures of around ten billion billion billion ($10^{28}$) degrees, would have been prevalent at about a hundred million billion billion billionth ($10^{-35}$) of a second after the big bang.

It seems reasonable to suppose that in this 'grand unification epoch', the strong nuclear force and electro-weak force would have been likewise indistinguishable, collapsing into a single 'electro-nuclear' force. All force carriers would have been identical and there would have been no mass, no electrical charge, no quark flavour (up, down) or colour (red, green, blue). Breaking this, even higher, symmetry required more Higgs fields, crystallizing at higher temperatures and so forcing a divide between quarks, electrons, and neutrinos and between the strong and electro-weak forces.

One of the first examples of such a grand unified theory (GUT) was developed by Glashow and Georgi in 1974.† This was based on the SU(5) symmetry group, which they declared to be the 'gauge group of the world'.[1] One consequence of the higher symmetry was that all elementary particles had simply become facets of each other. In Glashow and Georgi's theory, transformations between quarks and leptons had now

* More recent evaluation puts this energy somewhere in the region of two hundred thousand billion GeV.

† Although 'grand' and 'unified', GUTs do not seek to include the force of gravity. Theories that do so are often referred to as Theories of Everything, or TOEs.

become possible. This meant that a quark inside a proton could transform into a lepton. 'And then I realized that this made the proton, the basic building block of the atom, unstable,' Georgi said. 'At that point I became very depressed and went to bed.'[2]

As grand unification occurs only at energies that can never be realized in any earth-bound collider, it might be tempting to question the value of such theories. However, GUTs predict the existence of new particles which can in principle be revealed in collision experiments. And, although the grand unification epoch may have ended billions of years ago, it left a lasting imprint on the universe that we can observe today.

At least, this was the logic followed by young American postdoctoral physicist Alan Guth. He had confirmed that among the new particles predicted by GUTs was the *magnetic monopole*, a single unit of magnetic 'charge' equivalent to an isolated north or south pole. In May 1979 he had begun work with a fellow postdoc, Chinese-American Henry Tye, to determine the number of magnetic monopoles likely to have been produced in the big bang. Their mission was to explain why, if magnetic monopoles were indeed formed in the early universe, none are visible today.

Guth and Tye realized that they could suppress the formation of monopoles by changing the nature of the phase transition from grand unified to electro-weak epochs. This was a matter of tinkering with the properties of the Higgs fields involved. They discovered that the monopoles disappeared if, instead of a smooth phase transition or 'crystallization' at the transition temperature, the universe had instead undergone *supercooling*. In this scenario, the temperature falls so rapidly that the universe persists in its grand unified state well below the transition temperature.*

When in December 1979 Guth explored the wider effects of the onset of supercooling, he discovered that it predicted a period of extraordinary exponential expansion of space-time. Initially rather nonplussed by this result, he quickly realized that this explosive expansion could explain important features of the observable universe, in ways that the prevailing big bang cosmology could not. 'I do not remember ever trying to invent a name for this extraordinary phenomenon of exponential expansion,' Guth later wrote, 'but my diary shows that by the end of December I had begun to call it *inflation*.'[3]

Inflationary cosmology underwent some modifications largely as a result of further tinkering with the properties of the Higgs fields used to break the symmetry at the end of the grand unification epoch. These early theories predicted too much uniformity, implying a rather bland universe with no structure – no stars, planets, or galaxies. Cosmologists began to realize that the seeds of this observable structure had to come from quantum fluctuations in the early universe, amplified by inflation. But the properties of the Higgs fields required for this were incompatible with the Higgs fields of the Glashow–Georgi GUT.

By the early 1980s, experimental results were in any case confirming that the proton is more stable than Georgi and Glashow's theory implied.† No longer constrained by theories

* Liquid water can be supercooled to temperatures up to 40 degrees below freezing.

† These experiments involved searching for a single proton decay event from a large volume of protons shielded from cosmic rays. As Carlo Rubbia explained:

derived from particle physics, cosmologists were free to fit the observable universe by further tweaking the Higgs fields, which became collectively known as the *inflaton* field to emphasize its significance. Their predictions were borne out in spectacular fashion in April 1992 by results from the Cosmic Background Explorer (COBE) satellite, which mapped tiny variations in the temperature of the cosmic background radiation, the cold remnant of the hot radiation that had disengaged from matter about four hundred thousand years after the big bang.*

Brout and Englert, Higgs, Guralnik, Hagen, and Kibble had invented the Higgs field as a means to break the symmetries involved in Yang–Mills quantum field theories. Weinberg and Salam showed how the trick could be applied to electro-weak symmetry-breaking, and the technique was used to predict correctly the masses of the W and Z particles. The same trick had subsequently been used to break the symmetry of the electro-nuclear force. This trick had some surprising consequences, leading to the discovery of inflationary cosmology and the precise prediction of the large-scale structure of the universe.

The entirely theoretical notions of Higgs fields and the false vacuums they imply had become central to both the Standard

'. . . just put half a dozen graduate students a couple of miles underground to watch a large pool of water for five years.' Quoted in Woit, p. 104.

* These tiny temperature variations have since been measured in even more exquisite detail by the Wilkinson Microwave Anisotropy Probe (WMAP). Results reported in February 2003, March 2006, February 2008, and January 2010 have helped to confirm and refine the standard, so-called lambda-CDM (cold dark matter) model of the universe in which inflation plays a crucial part. According to the most recent WMAP data, the universe is 13.75±0.11 billion years old.

Model of particle physics and what would emerge as a Standard Model of big bang cosmology. Did these Higgs fields exist? There was only one way to find out.

The Higgs bosons of the grand unified Higgs fields possess huge masses and are simply out of reach of terrestrial colliders. However, although the mass of the Higgs boson of the original electro-weak Higgs field had proved hard to predict with any certainty, in the mid-1980s it was believed to be well within the grasp of the next generation of colliders.

American particle physicists were still smarting from being beaten to the discovery of the W and Z particles by their European rivals. A June 1983 *New York Times* editorial had declared 'Europe 3, US not even Z zero', and claimed that European physicists had now taken the lead in the race to discover the ultimate building blocks of nature.[4] The American physicists sought revenge. They were determined that the Higgs boson would be discovered at an American facility.

---

On 3 July 1983 Fermilab's Tevatron accelerator was turned on. Its six-kilometre ring reached its design energy of 512 GeV just twelve hours later. By colliding protons and anti-protons, it promised collision energies of 1 TeV. It had cost $120 million. ISABELLE, a new 400 GeV proton–proton collider under construction at Brookhaven, was now judged to be already obsolete. In July the project was cancelled by the US Department of Energy's High Energy Physics Advisory Panel.

Construction work was about to begin on CERN's LEP collider, to be housed in a 27-kilometre ring almost 600 feet beneath the French–Swiss border, which it would cross in four places. This would become the largest civil engineering project in Europe. But the LEP was intended as a W and Z particle factory, to be used for refining our understanding of the new particles and searching for the missing top quark. It was not a Higgs-hunter.

The Tevatron might provide opportunities to glimpse the Higgs boson, but there could be no guarantees. It was time to think big. Lederman had earlier proposed a giant leap forward – a super-massive proton–proton collider based on the use of superconducting magnets and capable of collision energies up to 40 TeV. He had called it the 'Desertron', because it would need to be built in a flat expanse of desert, and because it would be the only machine capable of crossing the 'energy desert', the energy gulf predicted by GUTs to be devoid of interesting new physics. The Desertron became the Very Big Accelerator (VBA). Having cancelled ISABELLE, the Advisory Panel now urged priority for the VBA, which was swiftly renamed the Superconducting Supercollider (SSC).

The design for the SSC was completed by the end of 1986. It came with a price tag of $4.4 billion, firmly propelling it into the big league of American science projects and requiring presidential approval. Lederman was asked to provide a short, 10-minute video about the project for President Ronald Reagan to review. He used the opportunity to appeal to Reagan's frontier spirit, drawing a direct analogy between the exploration of the unchartered areas of particle physics with the exploration of the American West.

The formal case for the SSC was put before Reagan and his Cabinet during a presentation at the White House in January

1987. Arguments for and against the investment bounced back and forth. Reagan's budget director argued that approval would achieve little more than make a bunch of physicists very happy. Reagan replied that this was something he probably should consider, as he had made his own physics schoolteacher very unhappy.

As the arguments subsided, attention turned to Reagan for a final decision. Reagan read out a passage by American writer Jack London: 'I would rather that my spark should burn out in a brilliant blaze than it should be stifled in dry rot.'[5] He explained that these words had once been quoted to the quarterback Ken 'Snake' Stabler. Stabler had steered the Oakland Raiders to a Super Bowl victory in 1977 and was famous for his passing accuracy and his 'Ghost to the Post', a 42-yard pass to Dave Casper (the 'Ghost') which set up an equalizing field goal in the dying seconds of an AFC playoff game against Baltimore Colts. The goal sent the game into overtime and the Raiders ran out eventual winners.

Stabler had interpreted the Jack London quote in the context of his own approach to American football. 'Throw deep,' Stabler had said.[6] In the face of adversity, it is better to adopt the riskier strategy and burn out in a brilliant blaze.

Reagan, a stalwart of American B-movies before entering politics in 1964, had acquired the nickname 'the Gipper' after appearing as college footballer George Gipp in the 1940 film *Knute Rockne, All American*. Gipp had died of a throat infection at the age of 25, and the film contains the famous quote: 'And the last thing George said to me, "Rock," he said, "sometime when the team is up against it and the breaks are beating the boys, tell them to go out there with all they've got and win just one for the Gipper." '[7]

There can be little doubt that Reagan found deep psychological resonance with the concept of the SSC. Already bedazzled by the promise that science could provide America with a last line of defence in the form of the Strategic Defense Initiative (SDI, also known as 'Star Wars'), he was now more than willing to go out there with all they've got in the interests of American scientific leadership. The Gipper was ready to throw deep.

---

The project had won approval, but was nevertheless beset by doubts. In its pitch the Department of Energy had explained how the SSC would become an international endeavour, supported by financial contributions from other countries. But the rhetoric from American physicists undermined this intention. Why would other countries support a project that was overtly designed to restore American leadership in high-energy physics? Europe was in any case firmly committed to CERN. Not surprisingly, the SSC attracted little interest from overseas.

Resentment had also built up within the American physics community, and this now spilled over into confrontation. With such a high cost, just what exactly was being sacrificed in the search for the Higgs boson? There were many other, individually much less costly projects that were much more likely to provide potentially valuable technological advances. American physics budgets couldn't fund all these and the SSC, and these projects now appeared to be at considerable risk. Was high-energy physics really a thousand times more valuable than other scientific fields?

'Big science' became a pejorative term.

Congressional and Senate support for the SSC was maintained for as long as the location of the new collider was unknown. The National Academies of Science and Engineering received 43 proposals from 25 different states. The Texas government established a commission, and it promised $1 billion funding if the SSC would be built in its territory. It might have made more sense to build the new collider at Fermilab, where much of the infrastructure and many of the physicists who would be needed were already established. But, in November 1988, the National Academies decided that the SSC would be built in a Cretaceous period geological formation called the Austin Chalk, deep beneath the Texan prairie in former cotton-rich Ellis County.

Reagan's vice-president, Texan George Bush, had succeeded him as president just two days before the announcement. There was no suggestion of bias in the National Academies' decision, but Bush became a strong supporter. However, now that the site location was known, support from other congressmen and senators began to evaporate.

The physicists had to battle continuously to wrestle funds from Congress, and were called to testify for the project every time it came up for review. In the meantime, budget estimates mushroomed as engineers began to understand more clearly the implications of constructing a huge ring of superconducting magnets. By the time funds were released to begin construction in 1990, budget estimates had almost doubled to $8 billion.

Test holes were drilled into the Austin Chalk and some of the infrastructure was built near Waxahachie, on part of a 17-thousand-acre site that had been reserved for the project by the Texan government. Laboratories were constructed for the development and testing of the magnets. Large structures were assembled to house the refrigeration units required to produce and circulate the liquid helium needed to keep the magnets at their superconducting temperature.

Two detector collaborations were formed. The Solenoidal Detector Collaboration (SDC) would consist of a thousand physicists and engineers from more than a hundred different institutions around the world. This would be a general-purpose detector costing $500 million. It was hoped that this would begin logging data before the end of 1999. The Gammas, Electrons, and Muons (GEM) group would be similar in size and would compete with the SDC.

Many physicists took a gamble and either organized a period of leave from their current jobs or quit their jobs altogether and relocated to join the SSC project. In all, about two thousand people gathered in and around Waxahachie. To an outsider unfamiliar with SSC politics, all this activity must have appeared rather reassuring. Laboratories were being built, holes were being drilled, and people were gathering in large numbers.

But there were other signs that were rather more ominous. The American administration was struggling with an already large and growing budget deficit. President Bush returned from a visit to Japan in January 1992 empty-handed: the Japanese insisted that the SSC was not an international project and, as such, they wouldn't support it. The noise about 'big science' was rising to a crescendo. In June the House of Representatives voted in favour of an amendment to the federal budget that would have shut the SSC project down. The project survived through the intervention of the Senate.

The gloom that was starting to gather around the project was reflected in Weinberg's popular book *Dreams of a Final Theory*, which was published in 1993. He wrote:[8]

> Despite all the building and drilling, I knew that funding for the project might yet be stopped. I could imagine that the test holes might be filled in and the Magnet Building left empty, with only a few farmers' fading memories to testify that a great scientific laboratory had ever been planned for Ellis County. Perhaps I was under the spell of [Thomas] Huxley's Victorian optimism, but I could not believe that this would happen, or that in our time the search for the final laws of nature would be abandoned.

In Lederman's rather more quixotic book, *The God Particle*, published in the same year, he is rudely awoken from a dream in which he has been chatting amiably to the Greek philosopher Democritus:[9]

> 'Shit.' I was back home, groggily lifting my head off my papers. I noticed one photocopy of a news headline: Congessional Funding for the Super Collider in Doubt. My computer modem was beeping, and an E-mail message was 'inviting' me to Washington for a Senate hearing on the SSC.

Bill Clinton won the November 1992 presidential election, beating George Bush and Independent Texas businessman Ross Perot. The following June, SSC budget estimates had grown to $11 billion and the House of Representatives once again voted against the project. As Raphael Kasper, the SSC associate director, remarked: 'Voting against the SSC became at some point a symbol of fiscal responsibility. Here was an expensive project that you could vote against.'[10]

Clinton was generally encouraging, but less committed to the project than Reagan and Bush had been. Competition now loomed in the form of a $25 billion programme to build the International Space Station, a project that would also be based in Texas, at NASA's Johnson Space Center in Houston.

In September 1993 Weinberg, Richter, and Lederman made a last-ditch attempt to shore up support for the SSC. The British physicist Stephen Hawking sent a message of support on video. To no avail.

In October the House of Representatives voted narrowly (by one vote) in favour of the International Space Station. The next day the House voted two-to-one against the SSC. This time there would be no reprieve. Funding was allocated to mothball the facilities that had already been built. About 23 kilometres of tunnel had been excavated and $2 billion had been spent (see Figure 19), but no amount of Victorian optimism could now keep the project alive. The SSC was dead.

Pulitzer prize-winning author Herman Wouk wrote a novel based on the SSC experience. His author's note at the beginning of *A Hole in Texas* says:[11]

Ever since coming up with the atomic and hydrogen bombs, [particle physicists] had been the pampered darlings of Congress. But all that suddenly and rudely ended. The sole residue of their miscarried quest for the Higgs boson was a hole in Texas, an enormous abandoned hole.

It's still there.

**FIGURE 19** When the SSC project was cancelled by Congress in October 1993, $2 billion had been spent and 23 kilometres of tunnel had been excavated beneath the Texas prairie.

*Source:* SSC Scientific and Technical Electronic Repository

On 16 December 1994, a little over a year after the SSC was cancelled, CERN's member states voted to allocate $15 billion over twenty years to upgrade the LEP at the end of its useful life and turn it into a proton–proton collider. The idea for the Large Hadron Collider (LHC) had first been discussed over ten years previously, at a CERN workshop in Lausanne, Switzerland, in March 1984. It would produce collision energies up to 14 TeV, less than half the maximum energy of the SSC but more than enough to find the Higgs.

Rubbia declared that CERN would 'pave the LEP tunnel with superconducting magnets.'[12]

# 9

# A Fantastic Moment

*In which the Higgs boson is explained in terms that a British politician can understand, hints of the Higgs are found at CERN, the Large Hadron Collider is switched on, and then blows up*

The SSC had been a huge gamble, and the physicists had lost. The rumblings of discontent that had led eventually to the cancellation of the American project had begun to surface in Europe. CERN benefitted from the fact that no one nation was responsible for its funding. But individual member states' grumbling about the size of their subscription could yet translate to a decision to withdraw support. In April 1993, just six months before the House of Representatives decided finally to cancel the SSC, UK science minister William Waldegrave was issuing a challenge to the British community of high-energy physicists.

Waldegrave's challenge presaged a significant shift in science policy by Prime Minister John Major's Conservative government. A government white paper, which was to be published the following month, sought to shift the emphasis of science policy towards innovation, with the ultimate aim to improve wealth creation and the quality of life of British citizens. In other words, the purpose of British science was to serve the interests of the British economy, to the benefit of 'UK plc'. The organization of government support for science and technology in Britain was to be completely overhauled.

The signs were ominous. Britain was still recovering from the global recession triggered by the stock market crash in October 1987, and could hardly afford its £55 million annual contribution to CERN. Whilst physicists could point to many spin-off developments from CERN, such as the project to join hypertext with the internet which led to the invention of the world wide web by Tim Berners-Lee in 1990, it was perhaps difficult to explain how the discovery of the Higgs boson would directly improve wealth creation and the quality of life of British people.

Fortunately, the physicists were not yet being asked to provide this kind of justification. But Waldegrave made it quite clear that they had to get an awful lot better at explaining precisely what it was they were trying to do.

Just what was this thing called the Higgs boson? Why was it so important that billions of dollars were needed just to find it? 'If you can help me understand that, I stand a better chance of helping you to get the money to find it,' Waldegrave told the audience at an annual conference of Britain's Institute of Physics.[1] He told them that if anyone could explain what all the fuss was about, in plain English, on one sheet of paper, then he would reward that person with a bottle of vintage champagne.

Of course, the fuss was all about the central role that the Higgs field had come to play in the structure of the Standard

Model. Without the Higgs field there could be no electro-weak symmetry-breaking.* Without symmetry-breaking the W and Z particles would be massless, like the photon, and the electro-weak force would still be unified. Without interactions between elementary particles and the Higgs field, there would be no mass: no material substance, no stars, no planets, no life. And the direct evidence for the existence of this field could only come from finding its field particle, the Higgs boson. Find the Higgs boson, and suddenly we would understand a lot more about the true nature of the material world.

Explaining the Higgs mechanism in terms that a politician could understand demanded a simple analogy. David Miller, a professor of particle physics and astronomy at University College London, believed he had found just such an analogy. With a few cosmetic changes, he thought he could bring the explanation to life by trading on Waldegrave's own experiences of a singular personality that had until recently dominated British politics: former Prime Minister Margaret Thatcher. He wrote:[2]

> Imagine a cocktail party of political party workers who are uniformly distributed across the floor, all talking to their nearest neighbours. The ex-Prime Minister enters and crosses the room. All of the workers in her neighbourhood

* Strictly speaking, this is not quite true. A class of 'technicolour' theories introduce new extra-strong forces which can also drive electro-weak symmetry-breaking. These theories can also account for the masses of the W and Z particles, but struggle to predict the quark masses correctly. For this reason, the Higgs mechanism is more favoured. Steven Weinberg, personal communication, 24 February 2011.

> are strongly attracted to her and cluster round her. As she moves she attracts the people she comes close to, while the ones she has left return to their even spacing. Because of the knot of people always clustered around her she acquires a greater mass than normal, that is, she has more momentum for the same speed of movement across the room. Once moving she is harder to stop, and once stopped she is harder to get moving again because the clustering process has to be restarted. In three dimensions, and with the complications of relativity, this is the Higgs mechanism.
>
> In order to give particles mass, a background field is invented which becomes locally distorted whenever a particle moves through it. The distortion – the clustering of the field around the particle – generates the particle's mass. The idea comes directly from the physics of solids. Instead of a field spread throughout all space a solid contains a lattice of positively charged crystal atoms. When an electron moves through the lattice the atoms are attracted to it, causing the electron's effective mass to be as much as 40 times bigger than the mass of a free electron.
>
> The postulated Higgs field in the vacuum is a sort of hypothetical lattice which fills our universe. We need it because otherwise we cannot explain why the Z and W particles which carry the weak interactions are so heavy while the photon which carries electromagnetic forces is massless.

This describes the mechanism by which massless elementary particles (represented in the analogy by Thatcher) interact with the Higgs field (the uniform distribution of party workers) and so gain mass, as shown in Figure 20. To explain the Higgs boson, Miller continued:

> Now consider a rumour passing through our room full of uniformly spread political workers. Those near the door hear of it first and cluster together to get the details, then they turn and move closer to their next neighbours who want to know about it too. A wave of clustering passes through the room. It may spread out to all the corners, or it may form a compact bunch which carries the news along a line of workers from the door to some dignitary at the other side of the room. Since the information is carried by clusters of people, and since it was clustering which gave extra mass to the ex-Prime Minister, then the rumour-carrying clusters also have mass.
>
> The Higgs boson is predicted to be just such a clustering in the Higgs field. We will find it much easier to believe that the field exists, and that the mechanism for giving other particles mass is true, if we actually see the Higgs particle itself. Again, there are analogies in the physics of solids. A crystal lattice can carry waves of clustering without needing an electron to move and attract the atoms. These waves can behave as if they are particles. They are called phonons, and they too are bosons. There could be a Higgs mechanism, and a Higgs field throughout our universe, without there being a Higgs boson. The next generation of colliders will sort this out.

This is illustrated in Figure 21.

**FIGURE 20** The explanation of the Higgs mechanism used by David Miller in his winning entry. As Margaret Thatcher makes her way through the 'field' of party workers, the field clusters around her and her progress is slowed. This is the equivalent of gaining mass.

*Source:* © copyright CERN

**FIGURE 21** The Higgs boson is like a softly spoken rumour that makes its way through the 'field' of party workers. As the field clusters together to hear the rumour, a localised 'particle' is formed which then makes its way around the room.

*Source:* © copyright CERN

Waldegrave received 117 entries, in itself indicative of the importance of the physicists' quest. Five winners were selected, but Miller's entry was judged by the physics community to be the best. Miller duly collected his bottle of Veuve Clicquot, although it seems he did not get to taste it. 'My wife, my sister-in-law and my son's girlfriend drank the champagne,' he explained.[3]

Despite its straightened circumstances, the British government continued to honour its commitment to CERN.*

---

With the hunt for the Higgs boson temporarily suspended, there were still a few other Standard Model particles to be found. The discovery of the top quark was eventually announced at Fermilab on 2 March 1995, by two competing research teams each consisting of about 400 physicists. It was identified through its decay products. Energetic protons and anti-protons collide to produce a top–anti-top pair. Each of these particles decays into a bottom quark and a W particle. One W particle decays into a muon and a muon anti-neutrino. The other decays into an up- and a down-quark. The end-result is a collision which produces a muon, muon anti-neutrino, and four quark jets. The mass of the top quark was found to be an astonishing 175 GeV, almost 40 times larger

* To put this in perspective, in 2011 Britain's contribution to the CERN budget was 15 per cent, or £109 million ($174 million), less than £2 per year for every UK citizen. 'That's literally peanuts,' said ATLAS physicist and TV presenter Brian Cox, 'In fact, we spend more here on peanuts than we do on the LHC.' (*Sunday Times*, 27 February 2011).

than the mass of its third-generation partner, the bottom quark.

Aside from the Higgs boson, the only other particle that remained to be discovered was the tau neutrino. Its discovery was announced at Fermilab five years later, on 20 July 2000. It was now possible to map the sequence of weak-force interactions that change one flavour of quark into another, Figure 22.

There was still some hope that the Tevatron or the LEP might find the Higgs boson, and these machines were now pushed to their limits. The problem was that the mass of the Higgs boson could not be predicted with any accuracy. Unlike with the search for the W and Z particles, the physicists didn't quite know where to look.

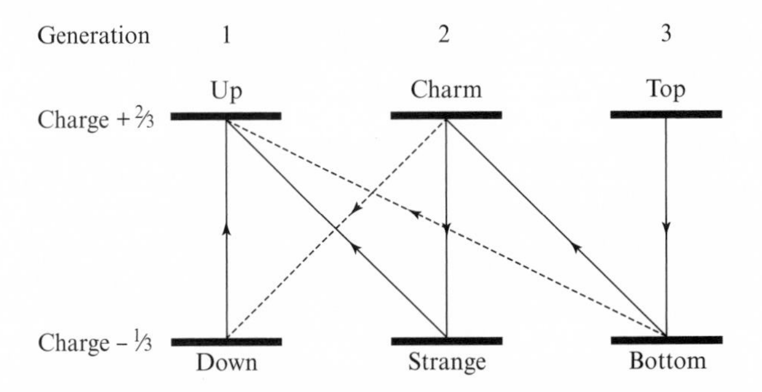


FIGURE 22 The dominant 'flavour-changing' weak force decays involving quarks are down→up, strange→up, charm→strange, bottom→charm and top→bottom. Two less probable decay paths are also shown (dashed lines): charm→down and bottom→up. Upward transitions involve the emission of a $W^-$ particle which decays into a lepton (such as an electron) and its corresponding anti-neutrino. Downward transitions involve the emission of a $W^+$ particle which decays into an anti-lepton (such as a positron) and its corresponding neutrino.

The collective understanding was that it would have a mass of the order 100–250 GeV. It would be detected through its decay channels, thought to involve production of bottom–anti-bottom pairs in association with top and bottom quarks, two high-energy photons, pairs of Z particles that would in turn decay into four leptons (electrons, muons, and neutrinos), pairs of W particles, and pairs of tau leptons.

The LEP had proved to be a powerful and versatile machine but had come to the end of its useful life and was due to be decommissioned in September 2000. In a last-ditch attempt to find the Higgs, CERN physicists now pushed the machine well beyond its limits. It had achieved its design beam energy of 45 GeV (producing electron–positron collision energies of 90 GeV) in August 1989. Various upgrades had lifted the collision energy to 170 GeV, giving the capability to generate pairs of W particles. In the summer of 2000, further modifications pushed collision energies above 200 GeV.

On 15 June 2000, CERN physicist Nikos Konstantinidis studied an event that had been recorded the previous day by the Aleph detector.* It featured four quark jets, two of which had come from the decay of a Z particle. The other two jets appeared to have come from the decay of a heavier particle, with a mass of the order of 114 GeV.

It looked to all the world like a Higgs boson.

Of course, a single event did not constitute a discovery, but it was quickly followed by two more events recorded by Aleph and two events recorded by a second detector collaboration,

* Aleph stands for Apparatus for LEP Physics.

called Delphi.[†] This was still insufficient to claim a discovery, but enough to persuade CERN's Director-General, Luciano Maiani, to stay LEP's execution until 2 November. When a third detector, called L3, recorded a different kind of event, which appeared to involve the decay of a Higgs into a Z particle which then decays into two neutrinos, it seemed that CERN was on the threshold of one of the most important discoveries in high-energy physics since the invention of the Higgs boson in 1964.

The CERN physicists now bid to keep the LEP running for another six months. Maiani seemed to be inclined to agree to this request but, after much soul-searching in a series of meetings with his senior research scientists, he eventually concluded that the evidence was insufficient to justify a potential delay to the construction of the LHC. There could be no managed transition, no graceful switching from the LEP to the LHC over an extended period of time. To build the LHC, the tunnel housing the LEP would have to be completely gutted. Maiani felt he had no choice but to close the LEP down. The CERN community learnt of the decision through a press release.

Many physicists were convinced that they were close to making a momentous discovery, and the way that Maiani had handled the situation left a bitter taste. However, when the collision events had been subjected to further scrutiny, the likelihood that these were really tell-tale signals of the Higgs boson was even further reduced. 'I understand the frustration and sadness of those who feel that they had the Higgs boson

[†] Delphi stands for Detector with Lepton, Photon, and Hadron Identification.

within their grasp,' wrote Maiani in February 2001, 'and fear that it may be years before their work can be confirmed.'[4]

All that could be concluded was that the Higgs boson must be heavier than 114.4 GeV, with a mass likely to be of the order of 115.6 GeV.

---

With the discoveries of the top quark and the tau neutrino, the collection of elementary particles that make up the Standard Model was virtually complete. Physicists faced the unprecedented situation that there were now no experimental data that did not conform to theoretical predictions. There was, nevertheless, much for the theorists to do.

The Standard Model's deep flaws had been painfully apparent from the very moment of its inception. The model has to accommodate a rather alarming number of 'fundamental' or 'elementary' particles. These particles are connected together in a framework that requires 20 parameters that cannot be derived from theory but must be measured. Of these 20 parameters, twelve are required to specify the masses of the quarks and leptons and three are required to specify the strengths of the forces between them.

Then there is the problem with the mass of the Higgs boson itself. The Higgs acquires mass through so-called 'loop corrections', which take account of its interactions with virtual particles. Loop corrections involving heavier particles such as a virtual top quark give the Higgs much more mass than it can afford to have if it is to break the electro-weak symmetry in the way that is required. The upshot is that the weak force is predicted to be a lot weaker than it really is. This is known as the 'hierarchy problem'.

And, despite Glashow, Weinberg, and Salam's ultimately successful combination of the weak and electromagnetic forces, the SU(3)×SU(2)×U(1) structure of Yang–Mills field theories that makes up the Standard Model is far from being a fully unified theory of particle forces.

Lacking guidance from experiment, the theorists had no choice but to be guided by aesthetics, following their instincts in the search for theories that could transcend the Standard Model and explain the laws of nature at an even more fundamental level.

In addition to grand unified theories of the Georgi–Glashow type, another approach to unification emerged in the early 1970s from theorists in the Soviet Union and was independently rediscovered in 1973 by CERN physicists Julius Wess and Bruno Zumino. This is called supersymmetry, often reduced to the acronym SUSY. There are many varieties of supersymmetric theories but one of the simplest – first proposed in 1981 and called the Minimal Supersymmetric Standard Model (MSSM) – features 'super-multiplets' which connect matter particles (fermions) with the bosons that carry forces between them.

In supersymmetric theories, the equations are invariant to the exchange of fermions for bosons, and vice versa. The very different properties and behaviours of fermions and bosons in the physics we observe today must then be the result of breaking or hiding this supersymmetry.

One consequence of this higher supersymmetry is the proliferation of more particles. For every fermion, the theory predicts a corresponding supersymmetric fermion (called a sfermion), which is actually a boson. This means that for every particle in the Standard Model, the theory requires a massive supersymmetric partner with a spin different by ½. The partner of the electron is called the selectron (a shortening of scalar-electron). Each quark is partnered by a corresponding squark.

Likewise, for every boson in the Standard Model, there is a corresponding supersymmetric boson, called a bosino, which is actually a fermion. Supersymmetric partners of the photon, W, and Z particles are the photino, wino, and zino.

One of the advantages of the MSSM is that it resolves the problem of the mass of the Higgs boson. In the MSSM, the loop corrections that lead the Higgs mass to inflate are cancelled by negative corrections resulting from interactions involving virtual super-symmetric particles. For example, the contribution to the Higgs mass arising from interactions with a virtual top quark is cancelled by interactions involving a virtual stop squark. This cancellation stabilizes the Higgs mass and hence the strength of the weak force. To make this mechanism work, the MSSM actually needs five Higgs particles, each with a different mass. Three of these particles are neutral and two carry electric charge.

The MSSM also irons out another wrinkle in the Standard Model. As Weinberg, Georgi, and Quinn had shown in 1974, the strengths of the Standard Model strong, weak, and electromagnetic forces become near-equal at high energies. But they do not become precisely equal, as might be expected in a field theory of a fully unified electro-nuclear force. In the

MSSM, the strengths of the three particle forces are predicted to converge on a single point (see Figure 23).

Supersymmetry may also resolve a long-standing problem in cosmology. In 1934, the Swiss astronomer Fritz Zwicky discovered that the average mass of galaxies in the Coma Cluster, inferred from their gravitational effects, is not consistent with the average mass inferred from the galaxies' luminosity in the night sky. As much as 90 per cent of the mass required to explain gravitational effects appeared to be 'missing', or invisible. This missing mass was called 'dark matter'.

This problem was not confined to a single cluster of galaxies. Dark matter is a central component of the current Standard Model of big bang cosmology, the lambda-CDM model. Successive observations of the cosmic microwave background radiation by the COBE and, more recently, WMAP, satellites suggest that dark matter constitutes about 22 per cent of the mass-energy of the universe. About 73 per cent is 'dark energy', associated with an all-pervasive vacuum energy field, leaving the 'visible' matter of the universe: stars, neutrinos, and heavy elements – everything we are and everything we can see – to account for less than five per cent.

Supersymmetry predicts super-particles that are not affected by either the strong or electromagnetic forces. Super-particles, such as neutralinos, are therefore candidates for so-called weakly interacting massive particles, or WIMPs, which are thought to constitute a significant proportion of dark matter.*

* Neutralinos are formed from combinations of photinos, zinos, and neutral higgsinos. See Kane, p. 158.

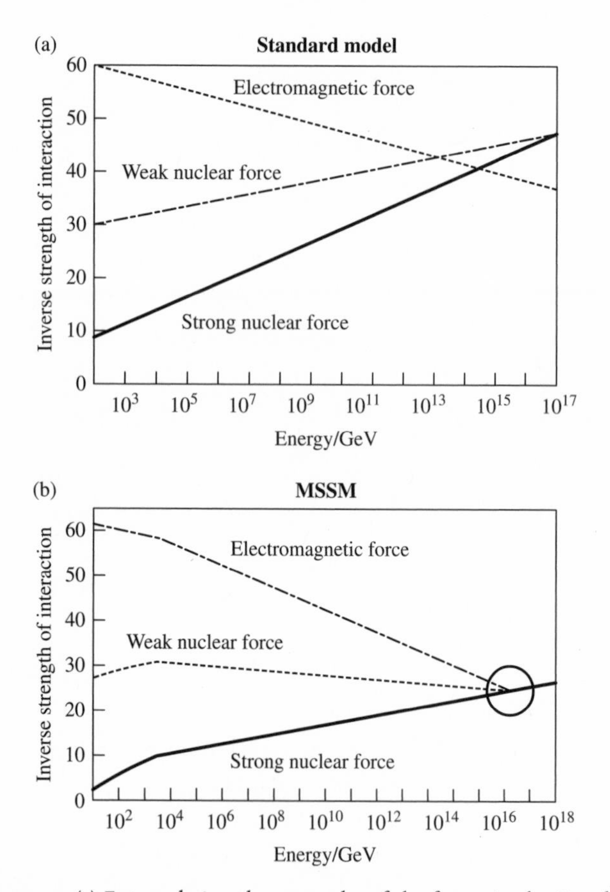


**FIGURE 23** (a) Extrapolating the strengths of the forces in the Standard Model implies an energy (and a time after the big bang) at which the forces have the same strength and are unified. However, the forces do not quite converge on a single point. (b) In the Minimum Supersymmetric Standard Model (MSSM) the additional quantum fields change this extrapolation, and the forces more nearly converge.

The existence of a host of supersymmetric particles may seem fantastic, but the history of particle physics is littered with fantastic discoveries based on theoretical predictions that many dismissed as absurd when they were made. If they do exist, some of the supersymmetric particles are anticipated to make their appearance at the TeV energy scale.

As the LHC began to take shape over 500 feet beneath French and Swiss soil at the beginning of a new millennium, it was obvious that its purpose was much more than finding the electro-weak Higgs boson, or indeed several Higgs bosons or supersymmetric particles as predicted by the MSSM. It was about pushing beyond the Standard Model; it was about our ability to understand what things are made of and how these things have shaped our universe.

---

Work began on dismantling the LEP in December 2000. Forty thousand tonnes of material had to be removed. The tunnel was completely emptied by November 2001, as surveyors began to mark the first of seven thousand locations for the components of the LHC.

There were inevitable delays. Maiani identified substantial cost over-runs in October 2001 and subsequent budget constraints pushed completion of the project back a further year, from 2006 to 2007. Just as the Americans had discovered during their abortive project to build the SSC, the novel technology of superconducting magnets tended to chew up rather more budget than had been anticipated.

Construction of the world's largest refrigeration system, capable of cooling the superconducting magnets to −271.4 °C, was completed in October 2006. The last of the LHC's 1746 superconducting magnets was installed in May 2007.

Although the LHC would be housed in the 27-kilometre tunnel that had been used for the LEP, further excavation was necessary to make room for new detector facilities. In the original planning for the LHC, four detector facilities were envisaged. These were A Toroidal LHC Apparatus (ATLAS), the Compact Muon Solenoid (CMS), A Large Ion Collider Experiment (ALICE), designed for the study of heavy ion (lead nuclei) collisions, and the Large Hadron Collider beauty (LHCb), a facility specifically designed to study bottom quark physics.

A further two, much smaller, detector facilities were subsequently added. TOTal Elastic and diffractive cross-section Measurement (TOTEM) is designed to make measurements of exquisitely high precision on protons and is installed near the point where protons collide in the centre of the CMS detector. Finally, the purpose of the Large Hadron Collider forward (LHCf) detector is to study particles generated in the 'forward' region of proton–proton collisions, almost directly in line with the colliding beams. It nestles alongside ATLAS and shares the beam intersection point.

The general-purpose ATLAS and CMS detectors would be involved in the hunt for the Higgs boson and other 'new physics' that might signal the existence of supersymmetric particles and resolve the riddle of dark matter. The ATLAS detector consists of a series of ever-larger concentric cylinders around the point at which the proton beams from the LHC intersect. The function of the inner detector is to track charged particles, enable their identification, and measure their momentum. The inner detector is surrounded by a large solenoid (coil-shaped) superconducting magnet which is used to bend the paths of the charged particles.

Sitting outside this are electromagnetic and hadronic calorimeters, which absorb charged particles, photons, and hadrons, and infer their energies from the particle showers they create. A muon spectrometer measures the momentum of highly penetrating muons which pass through the other detector elements. It makes use of a toroidal (doughnut-shaped) magnetic field created by large superconducting magnets formed into eight 'barrel loops' and two 'end caps'. These are the largest superconducting magnets in the world (see Figure 24).

ATLAS cannot detect neutrinos, and their presence must be inferred from the energy imbalance between colliding and detected particles. The detector must therefore be 'hermetic': no particles other than neutrinos can escape undetected.

The ATLAS detector is about 45 metres long and 25 metres high, about half as big as Notre Dame Cathedral in Paris. It weighs about seven thousand tonnes, equivalent to the Eiffel Tower or a hundred empty 747 jumbo jets. The ATLAS collaboration is led by Italian physicist Fabiola Gianotti and consists of three thousand physicists from more than 174 universities and laboratories in 38 different countries.

CMS has a different design but similar capabilities. The inner detector is a tracking system, made of silicon pixels and silicon strip detectors which measure the positions of charged particles allowing their tracks to be reconstructed. As in the ATLAS detector, electromagnetic and hadronic

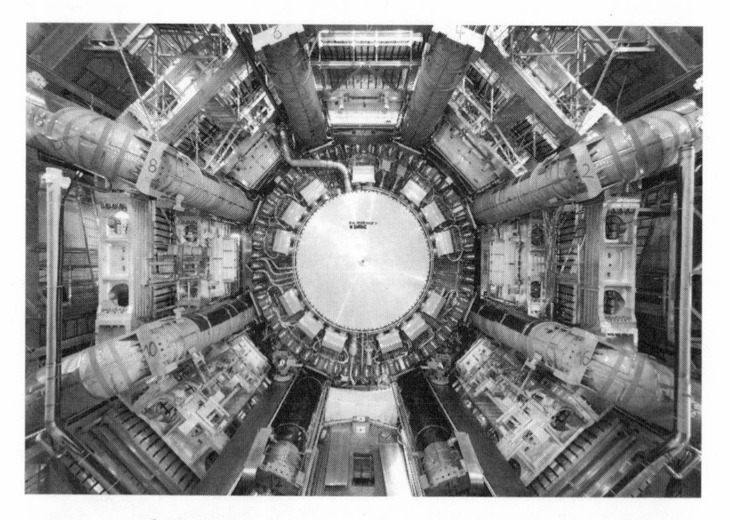

FIGURE 24 The ATLAS detector makes use of a toroidal (doughnut-shaped) magnetic field created by large superconducting magnets formed into eight barrel loops and two end caps. These are the largest superconducting magnets in the world.

*Source:* © copyright CERN

calorimeters measure the energies of charged particles, photons, and hadrons. A muon spectrometer captures data on muons which penetrate the calorimeters.

The CMS detector is 'compact', meaning that it uses a single, large solenoid superconducting magnet and so it is smaller than ATLAS. But it is still large: 21 metres long, 15 metres wide, and 15 metres high (see Figure 25). The CMS website proclaims that it sits in an underground cavern 'that could contain all the residents of Geneva; albeit not comfortably.'[5] The CMS collaboration is led by Italian physicist Guido Tonelli, and also includes three thousand physicists and engineers from 183 institutes in 38 countries.

FIGURE 25 Peter Higgs (on the left) visiting the CMS detector during the construction phase. He is pictured here with CMS spokesperson Tejinder Virdee.

*Source:* © copyright CERN

Work had begun in 1997 and 1998 on the construction of the ATLAS and CMS detector components and the excavation of the caverns that would house them. Assembly of both detectors was completed in early 2008.

By August 2008, all 27 kilometres of the LHC had been cooled to its operating temperature. The operation had required more than 10,000 tonnes of liquid nitrogen and 150 tonnes of liquid helium to completely fill the magnets.

The LHC was ready to be switched on.

'It's a fantastic moment,' Lyndon Evans, the LHC project manager, declared on 10 September 2008. 'We can now look forward to a new era of understanding about the origins and evolution of the universe.'[6]

Sadly, Evans' delight was not to last. The LHC was switched on at 10:28 am local time. Physicists crammed into the small control room cheered as a single flash of light appeared on a monitor, signifying that high-speed protons had been steered all the way around the machine's 27-kilometre ring at an operating temperature just two degrees above absolute zero. Though somewhat unspectacular (and something of an anticlimax for the estimated one billion people thought to have watched the moment on television), it represented the culmination of two decades of unstinting effort by armies of physicists, designers, engineers, and construction workers.

Another proton beam was sent around the ring in the opposite direction at 3 pm that day. Trouble began shortly afterwards. Just nine days later an electrical bus connection between two of the superconducting magnets short-circuited. Electricity arced, punching a hole in the magnets' helium enclosure. Helium gas leaked into sector 3–4 of the LHC tunnel, and in the subsequent explosion 53 magnets were damaged and the proton tubes were contaminated with soot.

There was no hope of repair before the scheduled winter shut-down, and a restart was tentatively fixed for spring 2009. But there were more problems, and at a meeting in Chamonix in February 2009 CERN managers took the decision to commission further work.

The restart date was pushed back.

# 10
# The Shakespeare Question

---

*In which the LHC performs better than anyone expected (except Lyn Evans), a year's data is gathered in a few months and the Higgs boson runs out of places to hide*

It was only at the beginning of September 2009, almost a year after it had first been switched on, that the last of the LHC's eight sectors began its cool-down procedure. All eight sectors were back at their operating temperature by the end of October, and the LHC was restarted in November. Despite the increased cost of electricity during the winter months, the collider was operated through the winter of 2009–10, primarily so that CERN physicists could stay ahead of their rivals at Fermilab's Tevatron, who had also produced tantalizing glimpses of the Higgs.

Through the first few months of 2010, protons running around the LHC in two rings travelling in opposite directions were accelerated to 3.5 TeV before being brought into head-on collision. The first 7 TeV collisions were recorded on 30 March. This collision energy was then maintained as the beam intensity and luminosity were gradually increased. Both ATLAS and CMS recorded events that could be ascribed to many familiar faces, as the panoply of Standard Model particles that had taken more than six decades to discover were registered in a matter of months. These included the neutral pion, first discovered in 1950, the eta, rho, and phi mesons (formed from various combinations of up-, down-, and strange-quarks), the J/ψ meson, the upsilon, and the W and Z bosons (see Figure 26). By July, new data on the top quark were being gathered.

On 8 July 2010, Italian physicist Tommaso Dorigo posted a blog entry reporting rumours that evidence for a light Higgs

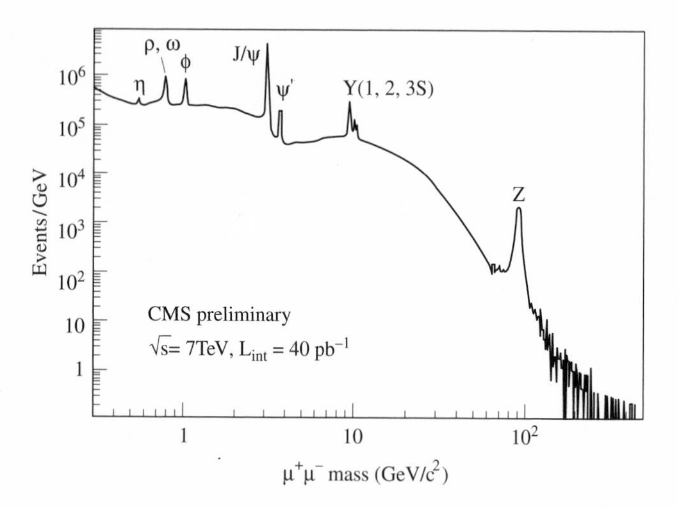


**FIGURE 26** In the first few months of 7 TeV operation in 2010, both the ATLAS and CMS collaborations recorded candidate events for the entire spectrum of known Standard Model particles. This diagram from the CMS collaboration shows evidence for the J/ψ, the upsilon (Υ, a meson formed from a bottom quark and its anti-particle) and the $Z^0$, revealed through production of muon–anti-muon pairs carrying different energies.
*Source:* Copyright CERN, for the benefit of the CMS Collaboration.

boson had been found at the Tevatron. The rumour spread rapidly around the internet and was picked up by the news media. It was almost immediately denied by Fermilab, in a 'tweet' which referred dismissively to 'rumours spread by one fame-seeking blogger'.[1] Dorigo subsequently sought to justify his rumour-mongering, arguing that '. . . keeping particle physics in the press with hints of possible discoveries that later die out is more important than speaking loud and clear once in ten years, when a groundbreaking discovery is actually really made, and keeping silent the rest of the time.'[2]

Right or wrong, the rumours were symptomatic of the growing rivalry between Fermilab and CERN and the growing sense of expectation that *something* might be discovered soon. Lederman had earlier admitted that watching as CERN announced any future discovery would leave him with mixed feelings: 'It would be a little like your mother-in-law driving off a cliff in your BMW,' he said.[3]

Dorigo's blog post had referred to rumours of 'three-sigma' evidence, a statistical measure reflecting the degree of confidence in the experimental data.* Three-sigma evidence would suggest a confidence level of 99.7% – in other words, a 0.3% chance that the data are in error. Although such confidence levels sound pretty convincing, to warrant declaration of a 'discovery', particle physicists actually demand five-sigma data, or confidence levels of 99.9999%.

Collision events leading to the production and decay of the Higgs boson were believed to be very rare, so building a five-sigma data set would require the recording of lots and lots

* There was obviously no such statistical measure for the rumour itself . . .

of candidate collisions. The particle beam luminosity was therefore key. The higher the luminosity, the greater the number of collisions within a fixed period, and the greater the number of potential candidate collisions.* In fact, the integrated luminosity (the sum of the luminosity over time), is directly related to the number of candidate collisions.

The integrated luminosity is reported in rather obscure units of inverse 'barns'. Physicists measure the rates of nuclear reactions in the form of 'cross-sections', reported in units of square centimetres. The cross-section can be thought to represent the size of a hypothetical two-dimensional 'window' through which the reaction occurs. The larger the window, the more likely the reaction. The more likely the reaction, the faster it will occur. The reported cross-sections have atomic dimensions, typically some number multiplied by $10^{-24}$ $cm^2$. The cross-sections for reactions involving atoms of uranium were found to be so large that one Manhattan project physicist quipped that they were as 'big as a barn'. The barn was subsequently introduced as a unit. A cross-section reported as some number times $10^{-24}$ $cm^2$ became some number of barns. A picobarn is a thousand billionth ($10^{-12}$) of a barn, or $10^{-36}$ $cm^2$. A femtobarn is a million billionth ($10^{-15}$) of a barn, or $10^{-39}$ $cm^2$.

At a CERN meeting in Evian, France, on 8 December 2010, Gianotti summarized the prospects for finding the Higgs and the nature of the race between the LHC and the Tevatron.

* The luminosity is a measure of the number of particles that can be squeezed into the collision point, and hence the number of *potential* collisions. Not all particles in the collision point will actually collide. Nevertheless, the luminosity gives the likelihood that a number of collisions will occur.

Simple statistics suggested that even with an integrated luminosity building to 10 inverse femtobarns (10 times $10^{15}$ inverse barns, or $10^{40}$ $cm^{-2}$) by the end of 2011, the Tevatron could do no better than report three-sigma evidence for the Higgs in certain restricted energy ranges. The more powerful LHC was in principle capable of generating three-sigma evidence with between 1 and 5 inverse femtobarns, depending on the Higgs mass.

On 17 January 2011, the US Department of Energy announced that it would not fund an extension to the Tevatron programme beyond the end of 2011. This decision did not signal the end of the race for the Higgs, but it did acknowledge the inevitable passing of custodianship of the frontier of high-energy physics from Fermilab to CERN.

The original operational plan for the LHC had included a prolonged shutdown in 2012, necessary to upgrade the proton beam energies to deliver the design collision energy of 14 TeV.* With the Higgs so tantalizingly close, in January 2011 CERN managers agreed to postpone the shutdown and continue to operate the LHC at a collision energy of 7 TeV through to December 2012. A potential upgrade to a collision energy of 8 TeV was judged to be too risky. Instead, ways to increase the beam luminosity would be implemented.

'If nature is kind to us and the Higgs particle has a mass within the current range of the LHC,' said CERN Director-General Rolf Heuer of the decision, 'we could have enough

* The shutdown was judged to be necessary in order to open up some 27,000 interconnections between the main superconducting magnets, repair them, and clamp them together so that they would support the higher currents necessary to deliver 7 TeV per beam.

data in 2011 to see hints, but not enough for a discovery. Running through 2012 will give us the data needed to turn such hints into discovery.'[4]

The stage was set.

---

Einstein's secretary Helen Dukas once asked if he could provide a simple explanation of relativity that she could use in response to the many queries she received from reporters. He thought about it for a while and suggested: 'An hour sitting with a pretty girl on a park bench passes like a minute, but a minute sitting on a hot stove seems like an hour.'[5]

Among the thousands of scientists involved in the collaborations at Fermilab and CERN, the tension and excitement was now palpable. There had been no particle discovery in over a decade. Nearly eleven years had passed since the Higgs had been 'glimpsed' by the LEP collider. And now the promise of new physics was desperately, agonisingly close. What was it? Six months? A year? Two years? This was definitely 'hot stove' territory.

It was perhaps inevitable that the dam would burst.

Columbia University mathematical physicist Peter Woit had maintained a blog on high-energy physics since the 2006 publication of his successful book *Not Even Wrong*, a critique of contemporary string theory. On 21 April 2011 he received an anonymous posting which contained the abstract of an internal ATLAS discussion paper. The paper claimed to have found four-sigma evidence for a Higgs boson with a mass of 115 GeV.

This was not a hoax. The paper was authored by a small team of ATLAS physicists at the University of Wisconsin-Madison under the leadership of Sau Lan Wu, who had been part of the Aleph collaboration that had 'glimpsed' the Higgs in 2000 towards the end of the LEP's lifetime. It was therefore no coincidence that Wu had gone back to look at the energy range where she believed she had seen those earlier hints.

There were two problems, however. The first was physical. The particle had been observed in the so-called di-photon mass distribution from a combined total of about 64 inverse picobarns of data gathered during 2010 and early 2011.

At an energy of 7 TeV, the proton-proton collisions occurring in the LHC actually involve quark-quark collisions and the fusion of gluons which, in theory, may produce Higgs bosons. The decay channels open to the Higgs depends on its mass. For a large-mass Higgs decay channels involving the production of two W particles or two Z particles would be available. But for a low-mass 115 GeV Higgs there is insufficient energy to reach these channels. Instead, the Higgs decays through alternative routes. One of these involves the production of two high-energy photons, a process written as $H \rightarrow \gamma\gamma$.

The problem was that the observed resonance was about 30 times larger than the Standard Model prediction for this particular decay channel.

Decay of the Higgs into two photons is dominated in the Standard Model by so-called W-boson 'loops' involving the production and subsequent annihilation of W bosons. The upshot is that this decay route is predicted to happen very infrequently, accounting for only about 0.2 per cent of all possible decay pathways. If this really was the Higgs, then its decay into two photons was for some reason being greatly enhanced. Other new particles, such as fourth or even fifth generation quarks and leptons, might have to be invoked to explain this.

The second problem concerned the status of the finding. The leaked document was an internal, so-called ATLAS-Communications or 'COM' note – designed for the rapid distribution of un-vetted and un-approved results for discussion within the collaboration. There was no sense in which this could be construed as an 'official' view from the ATLAS group. Subsequent review and re-analysis could eliminate this result entirely, long before any formal paper could be written.

The news of the leaked COM note hit the 'blogosphere' just before the long Easter weekend and, for a few days, the discussion was contained among high-energy physics bloggers and their followers. In 2009 Dorigo had predicted that news of the discovery of the Higgs would appear first in a blog post. He felt that his prediction had been validated, but he was nevertheless highly doubtful that this was the Higgs and offered $1000 against $500 that further data would reveal no new 115 GeV particle in the di-photon decay channel.

The story was picked up by the British mainstream media on Easter Sunday, 24 April. Jon Butterworth, an ATLAS physicist based at University College London, gave a balanced report to the UK's Channel 4 News. He said: 'What's happened here is a bunch of people have spent four nights without sleep. They've made some plots and got rather over-excited, [and] sent them in an internal note around the collaboration.

Which is fine. Everyone's excited out there but unfortunately it's leaked out. It's a very exciting place at the moment.'[6] The story was reported widely in the newspapers the following day.

In his blog for the *Guardian* newspaper, Butterworth expanded on this theme: 'Retaining a detached scientific approach is sometimes difficult. And if we can't always keep clear heads ourselves, it's not surprising people outside get excited too. This is why we have internal scrutiny, separate teams working on the same analysis, external peer review, repeat experiments, and so on.'[7]

Counter-rumours appeared shortly afterwards. A French high-energy physics blog claimed on 28 April that, on examining more data, ATLAS physicists were finding that the evidence for the Higgs was quickly disappearing. On 4 May *New Scientist* staff reporter David Shiga posted an online news item claiming to have seen a document leaked from the CMS collaboration which indicated that a search through their data had 'come up empty'.[8] Through such leaks the interested observer caught glimpses of the chaotic to-ing and fro-ing now going on within the ATLAS and CMS collaborations.

On 8 May, the ATLAS collaboration released an official update. Further analysis of a total of 132 inverse picobarns of data from 2010 and 2011 had indeed come up empty; the di-photon mass distribution showing no excess of events. In a subsequent blog posting, Butterworth explained that this null result has hardly surprising: even the Standard Model predictions suggested that there should be nothing to see yet, but that something could be expected 'soonish'. 'So stay interested in the di-photon mass spectrum,' he wrote, 'but wait for solid results before opening the champagne.'[9]

It seemed that we wouldn't have to wait too long. At midnight on 22 April, the LHC set a new world record for instantaneous luminosity, of $4.67 \times 10^{32}$ cm$^{-2}$ s$^{-1}$, or 467 inverse microbarns (millionths of a barn) per second. The engineer-in-charge that evening, Laurette Ponce, had visited CERN as a child and had joined the laboratory in 1999 to study for her PhD. 'I never imagined then that one day it would be me pressing the button to fill the LHC,' she said.[10]

As it was midnight, there were few in the CERN Control Room to witness the moment. Ponce shouted and danced, waving her arms in the air like a teenager.

This dramatic increase in luminosity had been achieved by injecting more and more proton bunches from the SPS into each beam circulating around the LHC. By 3 May, the peak luminosity had been increased further, to 880 inverse microbarns per second with 768 bunches per beam. Towards the end of May, a peak luminosity of 1260 inverse microbarns per second was recorded.

To put this in perspective, the cross-section for inelastic proton-proton collisions at 7 TeV is about 60 millibarns, or 0.06 barns. An instantaneous luminosity of 1260 inverse microbarns per second therefore implies $1260 \times 10^6 \times 0.06 =$ over 75 *million collisions per second*. If we take the cross-section for Higgs boson production at 7 TeV to be 9 picobarns,* then

* This is based on recommendations for collision energies of 7 TeV reported by the LHC Higgs Cross Section Working Group. The calculated cross sections for Higgs production from gluon-gluon fusion processes vary depending on the mass

this instantaneous luminosity implies $1260 \times 10^6 \times 9 \times 10^{-12} = 0.011$ Higgs bosons per second, or *one Higgs boson produced on average every 90 seconds.*

The furore over the leak had sparked interest in the process by which a 'blockbuster' result would be officially announced. James Gillies, CERN's Head of Communications, explained to *New Scientist* that any such result would first be discussed and agreed within the collaboration (ATLAS or CMS) that found it, before being communicated to the CERN Director-General. It would then be communicated to the second collaboration, so that the result could be corroborated. Then the heads of other laboratories and the individual member states that contribute to CERN's funding would be advised. The announcement would then take the form of a seminar organised at CERN. By this time, many thousands of people would be in the know. A further leak appeared not only very possible, but almost inevitable.

So, where would the dam break next?

By 17 June, the LHC had already delivered its milestone 1 inverse femtobarn of data to each of the detector collaborations, an objective that had been set for the whole of 2011. 'I don't think we set the targets too low,' Heuer explained in his mid-year talk to staff. 'I think we set the targets realistically, but not optimistically. And I must say for me, as a born optimist, the machine was running better than expected.'[11]

But for Lyn Evans, this was no real surprise. 'The LHC is working much better than anybody expected, except me,' he

of the Higgs, from about 18 picobarns at 115 GeV to around 3 picobarns at 250 GeV. The average in this Higgs mass range is about 9 picobarns.

declared. 'I'm very happy.'[12] Evans had joined CERN in 1969 and had been part of the LHC project from its inception at the Lausanne workshop in 1984. He had led the project since 1993. It had been an emotional journey.

With this much data now delivered to both ATLAS and CMS, expectations rose to an all-time high. This should be enough data to provide three-sigma evidence for a Higgs boson in the mass range 135–475 GeV. Or it should be enough data to exclude the Higgs with 95 per cent confidence from 120–530 GeV. Projecting forward to the end of 2012, it seemed that the matter would be definitively settled, one way or another.

'To my mind, the Shakespeare question for the Higgs – to be, or not to be – will be answered at the end of next year,' said Heuer.[13]

All attention now turned to the European Physical Society (EPS) high-energy physics conference in Grenoble, France, scheduled to start on 21 July.

---

The EPS meeting was to be the first opportunity for the ATLAS and CMS collaborations to share what each had found with over 1 inverse femtobarns of data. That the collaborations could present results literally within weeks of the data being gathered was testimony to the hard work and commitment of hundreds of physicists who had worked tirelessly – and with little sleep – on the analysis.

It became apparent that the Higgs boson (or bosons) – if such existed – was not about to be 'found', as such. Instead, possible Higgs mass ranges would first be eliminated from enquiries, narrowing the search to smaller and smaller mass ranges until, finally, the Higgs would be left with nowhere to hide.

The ATLAS collaboration could now exclude a Standard Model Higgs boson with a mass between 155–190 GeV and 295–450 GeV, with 95 per cent confidence. In itself, this was already a powerful result. Finding nothing across such a broad energy range put the cat among several different theoretical pigeons; most of them concerned with physics beyond the Standard Model.

But there was more. The ATLAS data also showed an excess of events above background between 120–145 GeV. This could be due to a number of things, such as errors in the analysis, fluctuations in the background of events that hadn't been properly anticipated or calculated, or systematic uncertainties in the detector. Or this could be the first intimation that something like a Standard Model Higgs boson, or perhaps even multiple Higgs bosons, were lurking in this energy range.

The excess was dominated by events which could be ascribed to two different Higgs decay channels. These involved a Higgs decaying to two W particles, and thence to two charged leptons and two neutrinos (written $H \rightarrow W^+W^- \rightarrow \ell^+\nu\ \ell^-\nu$)*, with a smaller contribution from the channel in which a Higgs decays to two $Z^0$ particles and thence to four charged leptons (written $H \rightarrow Z^0Z^0 \rightarrow \ell^+\ell^-\ell^+\ell^-$).† The former was expected to be one of

* The leptons and neutrinos are produced in combinations. For example, a $W^-$ particle will decay into an electron or a muon and a corresponding anti-neutrino, a $W^+$ particle will decay into a positron or anti-muon and a corresponding neutrino.

† Again, the leptons are produced in combinations: electrons with positrons, muons with anti-muons.

the most dominant decay channels for a Standard Model Higgs of sufficient mass but, of course, the neutrinos and anti-neutrinos so produced had to be inferred because they can't be detected, and it is notoriously difficult to distinguish genuine Higgs events from the background. Consequently, only a range of Higgs masses could be implied from data for this channel.

The second channel is much cleaner. In fact, this is the 'golden' channel, so called because it is almost completely free from background events and so gives a potentially very precise measure of the Higgs mass. It is also very rare, with only about one in every thousand Higgs bosons decaying this way.

The observed excess of events in the combined ATLAS data was just 2.8 standard deviations, or 2.8-sigma, above background. This was not quite three-sigma 'evidence' and far from the five-sigma needed to declare a discovery. It was, nevertheless, a strong hint. What had CMS found?

The CMS collaboration announced 95 per cent confidence limit exclusions over the ranges 149–206 GeV, much of the region 200–300 GeV and 300–440 GeV. The combined CMS data also showed an interesting excess of events between 120–145 GeV, with a statistical significance that had proved difficult to evaluate but which was slightly less than ATLAS had claimed.

This was electrifying. ATLAS and CMS, which before the conference had worked separately, secretively and in competition, had both found much the same thing.

There was still a very long way to go. After the presentations some members of the ATLAS and CMS collaborations gathered together to celebrate with champagne and discuss the next steps. A small working group would be convened to combine the results from the two collaborations and update them with new data to provide a more definitive assessment.

The LHC had continued to break its own records. On 30 July it reached a peak luminosity of 2030 inverse microbarns per second (more than 120 million proton-proton collisions per second). Despite some stability problems, by 7 August the collider had delivered more than 2 inverse femtobarns of data to both ATLAS and CMS. This was already twice the amount of data that had been analysed and shown at the EPS conference.

The combined and updated results would be ready in time for the next big summer conference, the Fifteenth International Symposium on Lepton-Photon Interactions at High Energies, scheduled to commence on 22 August at the Tata Institute in Mumbai, India.

It seemed that the answer to the Shakespeare question could be available within months.

---

Einstein once declared: 'The Lord is subtle, but he is not malicious'.* Although the next chapter in the saga of the Higgs search might not betray the existence of an overly malicious

* The phrase 'Raffiniert ist der Herr Gott. Aber Boshaft ist Er Nicht' is carved in stone above the fireplace in a room in Fine Hall, Princeton University, in memory of Einstein.

deity, as the events unfolded it would become reasonable to accuse the Lord of possessing a certain mischievous wit.

In the weeks before the Mumbai conference, rumours began to circulate in the blogosphere that the combined ATLAS and CMS data now spoke much *less* ambiguously of a Higgs boson with an energy around 135 GeV. It seemed that the combined ATLAS and CMS data suggested an excess of Higgs decay events with much greater than three-sigma significance. Expectations built in intensity. Although three-sigma evidence would not represent a 'discovery', it would be possible to judge from the confidence of the physicists closest to the results whether they believed that this indeed was 'it'.

I met with Peter Higgs on a wet Thursday afternoon in Edinburgh, a few days before the Mumbai conference was due to commence. Higgs had retired in 1996 but had remained in Edinburgh close to the University department where he had first become a lecturer in mathematical physics in 1960. He was now a sprightly 82 years old. We sat in a coffee shop with his colleague and friend Alan Walker, and talked about his experiences and his hopes for the near future.

Higgs had published the paper that was to bind him forever to the particle that bears his name in 1964.* He had waited 47 years for some kind of vindication. We talked about the prospects for the Mumbai conference, and the grounds we had for optimism that something momentous may be about to be reported. 'It's difficult for me now to connect with the person I was then [in 1964],' he explained, 'But I'm relieved it's

* Although the particle he predicted didn't become widely known as the Higgs boson until 1972.

coming to an end. It will be nice after all this time to be proved right.'[14]

Finding the Higgs boson would inevitably be rewarded with a Nobel Prize for the Higgs mechanism, and debates had raged about precisely who of those involved would be recognised by the Nobel Committee: Englert, Higgs, Guralnik, Hagen and Kibble.[†] We talked about the outbreak of publicity likely to surround a strongly positive announcement from Mumbai and any subsequent announcement from the Swedish Academy. The Press Office at Edinburgh University would be heavily involved. And if things got out of hand Higgs would simply unplug his phone and refuse to answer his doorbell.

But, it seemed, there would be no call for these extreme measures just yet. As the Mumbai conference got underway the following Monday, 22 August, James Gillies at CERN issued a press release. There was no mention of the combined ATLAS and CMS data that had been promised in Grenoble. When updated with another inverse femtobarn or more of collision data gathered in the time between the conferences, the excess events observed by both ATLAS and CMS in the low mass region around 135 GeV had actually *declined* in significance. 'Now, with additional data analysed, the significance of those fluctuations has slightly decreased,' declared the press release, rather solemly.[15]

It was hard not be disappointed. The hints that had surfaced in the results presented in Grenoble had become less

[†] Sadly, Robert Brout died in May 2011, after a long illness. The Nobel Prize is not awarded posthumously, and each Prize can only be shared by three individuals.

significant in results presented in Mumbai. The outstanding performance of the LHC in delivering more than two inverse femtobarns of data to each detector facility by August had helped build expectations that the 'Shakespeare question' might be answered sooner rather than later. The Lord had obviously decided to be mischievous – it wasn't going to be that easy.

Although data from more than 140 trillion proton-proton collisions had now been captured by each detector collaboration, the physicists were still wrestling with only a handful of excess events. And the statistics of the few can be prone to wild fluctuations. Small changes can make big differences.

For example, the statistics of tossing a coin seem very straightforward. We know that there's a 50:50 chance of getting heads or tails. However, if we look at only a few such tosses, we shouldn't be surprised if we see a sequence featuring an excess of heads or tails. This doesn't mean that the coin isn't 'true'. It simply means that we haven't observed enough tosses to give us a representative sample. As we gather more data, we would expect that any excess would gradually disappear.

The results presented in Mumbai did not yet mean that the Standard Model Higgs did not exist. There were still excess events at energies between 115–145 GeV, but this was an energy range that had always been acknowledged to be rather problematic for the LHC.

There was only one thing for it. We would have to be more patient and wait for even more data. Higgs had waited 47 years. Another few months wouldn't make much difference.

---

The LHC continued to perform better than expectations through the summer of 2011 and on into the autumn, reaching a peak luminosity of 3650 inverse microbarns per second. The proton run ended on 31 October, with each detector collaboration having amassed more than 5 inverse femtobarns of data from 350 trillion proton-proton collisions.

But these were vaguely troubling times. The experiences at Mumbai had undermined confidence. There had been no announcement on the Higgs from CERN since the Mumbai conference, and no announcement appeared to be forthcoming. The long-promised combined data from ATLAS and CMS was finally released, but this told us nothing new and referred only to the two inverse femtobarns of data that had been available in July. The combined data set was now more than five times as large.

A flurry of excitement was instead caused by the announcement, on 23 September 2011, that a group of physicists working at the OPERA experiment,* buried deep beneath the Gran Sasso mountain in the Appennine range in central Italy, were about to report the results of painstaking measurements of the speed of muon neutrinos generated at CERN, 730 kilometres away. The results suggested that the neutrinos were travelling through the earth and reaching their destination very slightly *faster* than the speed of light.

As the debate about faster-than-light neutrinos evolved, other CERN physicists were busy trying to explain how non-discovery of the Higgs still represented an important step

* OPERA stands for Oscillation Project with Emulsion-tRacking Apparatus and is a collaboration between CERN and the Laboratori Nazionali del Gran Sasso (LNGS).

forward for high-energy physics. It was certainly true that non-discovery would undermine the Standard Model and have theorists scrambling back to the drawing-board. But, with the best will in the world, finding nothing is simply not the same as finding *something*.

With the outlook rather gloomy, an announcement that the CERN Council would hold a meeting with member state representatives to discuss the latest developments in the search for the Higgs appeared largely underwhelming. The first day of the meeting, scheduled for 12 December 2011, would be closed. But public talks from Gianotti and Tonelli scheduled for the next day looked a little more promising. Was there something interesting to tell, after all?

The world's media gathered at CERN on Tuesday, 13 December. Journalists were no doubt somewhat bemused by the rather dry, technical presentations they witnessed, but the conclusions were nevertheless quite compelling.

Combining the data from several different possible Higgs decay channels, the ATLAS collaboration had observed an excess of events corresponding to 3.6-sigma above the predicted background for a Higgs boson with a mass of 126 GeV. CMS reported a combined excess of events with slightly lower statistical significance of 2.4-sigma for a Higgs with a mass around 124 GeV.

The physicists nevertheless urged caution. 'This excess may be due to a fluctuation,' said Gianotti, 'But it could also be something more interesting. We cannot conclude anything at this stage. We need more study and more data. Given the outstanding performance of the LHC this year, we will not need to wait long for enough data and can look forward to resolving this puzzle in 2012.'[16]

Heuer explained: '[The data provide] intriguing hints in several channels in two experiments, but please be prudent. We have not found it yet. We have not excluded it yet. Stay tuned for next year.'[17] Jon Butterworth told Britain's Channel 4 News: 'We're all rather excited because it looks very suggestive and, as Rolf Heuer said, it's turned up in a few different places at once. But we still need to roll the dice a few more times.'[18]

Higgs himself echoed this party line: 'Ah well, I won't be going home to open a bottle of whisky and drown my sorrows, but equally I am not going home to crack open a bottle of champagne either!'[19]

In a blog entry posted the same day, Dorigo declared the results to be 'firm evidence' for a standard model Higgs boson with a mass around 125 GeV.[20] There followed a short but intense war of words in the blogosphere as American theorist Matt Strassler adopted a more conservative view, arguing that Dorigo's use of the word 'firm' was unwarranted: 'If he had said "some preliminary evidence" he would have gotten away with it. As it is, it seems to me that he has crossed a line . . .'[21]

In truth, the physicists were collectively urging prudence, but many were individually ready to take a gamble, as Butterworth explained to me: 'We really do need data to be sure, but I would bet on this myself. [It] depends how much of a betting man you are.'[22]

At the very least, there were good grounds for optimism. With the LHC scheduled to re-start proton physics in April

2012, the focus of attention would once again turn to the big summer conferences.

---

The parameters for the next physics run at the LHC were decided at a workshop in Chamonix in February 2012. After a year of highly successful operation, the engineers were now much more confident about the machine's capabilities, and agreed to push the total proton-proton collision energy to 8 TeV. This higher energy could be expected to offer up to thirty per cent enhancement of the rate of Higgs production which, when the effects of increased backgrounds were taken into account, still resulted in a 10-15 per cent increase in sensitivity. The target was set to collect 15 inverse femtobarns of data at this higher collision energy during 2012. This would surely be enough data finally to end the search for the Higgs.

On 22 February, it was revealed that the OPERA results implying faster-than-light neutrinos were in error. A loose fibre-optic cable had caused a slight delay to the timing measurements, which had translated into a decrease in reported flight-time of the neutrinos of about 73 billionths of a second. When corrected, the measurements were entirely consistent with neutrinos travelling at light-speed.

To a large extent, this was an embarrassing conclusion to the saga, but physicists everywhere breathed a sigh of relief, secure in the knowledge that Einstein's special theory of relativity was safe. A couple of high-profile members of the OPERA collaboration resigned their positions. This was a rather sobering reminder (if one was needed) of what can happen when an elaborate physics experiment makes some very public announcements that are subsequently shown to be wrong.

Operations at the LHC were re-started on 12 March and the collision energy of 8 TeV was achieved eighteen days later. The proton physics run began in earnest in mid-April. Instantaneous luminosity built to a peak of 6760 inverse microbarns per second. Although data gathering was slowed slightly by some technical problems associated with the cryogenics, by the end of May the LHC was delivering an impressive 1 inverse femtobarns of data *per week* to each detector collaboration.

Strong momentum was now building for an announcement at the 36$^{th}$ International Conference on High Energy Physics (ICHEP), scheduled to commence on 4 July in Melbourne, Australia. By 10 June, the notional cut-off date beyond which it would not be possible to analyse further data in time to be presented at the conference, the LHC had delivered about 5 inverse femtobarns to both ATLAS and CMS, as much data as had been gathered through the whole of 2011.

Inevitably, rumours began to appear in the high-energy physics blogs. Peter Woit reported a rumour suggesting that strong hints of the Higgs were again being seen, with data from 2011 and about half the data available for 2012 showing an excess of events with 4-sigma significance in the $H \rightarrow \gamma\gamma$ channel. Speculation grew in intensity. All the signs were that both ATLAS and CMS might report data showing excesses just short of the 5-sigma needed to declare a discovery. If this were really the case, then there could be little doubt that combining the results from both collaborations would likely tip the conclusion in favour of something like the Higgs.

But would the collaborations take this step? If they didn't, then the matter would remain officially unresolved until yet more data had been obtained. This would leave the bloggers free to publish their own quite reasonable but definitely unofficial data combinations. The bloggers could find themselves possibly declaring a 'discovery' that couldn't be officially sanctioned. There was no precedent for this situation in the entire history of science.

And then, in a surprise move, CERN announced that it would hold a special seminar at the laboratory in Geneva on 4 July, as a 'curtain-raiser' to the ICHEP conference. The seminar would provide updates on the search for the Higgs from ATLAS and CMS, to be followed by a press conference. Higgs, Englert, Guralnik, Hagen and Kibble were all invited to attend.*

Surely, this was a sign that one or both detector collaborations had achieved the 5-sigma significance required to declare a discovery? The speculation intensified. Not to be outdone, Fermilab physicists reminded us that each of the two Tevatron collaborations, D0 and CDF, had accumulated almost ten inverse femtobarns of data at a lower collision energy. At a conference held in Moriond, France, in March physicists from Fermilab had revealed results suggesting a 2.2-sigma excess in the range 115-135 GeV, with emphasis on decays to two bottom quarks, a channel not easily observed at the LHC because of high background. In a subsequent seminar on 2 July, two days before the CERN announcement, Fermilab physicists declared

* Kibble had other commitments that day. Higgs, Englert, Guralnik and Hagen all attended the seminar.

that through improvements in their analysis they had pushed the significance to 2.9-sigma. Of course, this was insufficient to declare a discovery, but would certainly provide strong corroboration for any subsequent discovery announcement.

---

On 4 July I watched the live webcast from CERN from the comfort of my office, and tracked the audience reaction through blog entries posted live by Dorigo, who was present at the seminar.

Heuer declared this day to be special for several reasons. This was, after all, the opening event of an international physics conference, the first such conference to be opened via video link from a different continent.

First up was Joe Incandela, professor of physics at the University of California, Santa Barbara, acting as spokesperson for CMS. He seemed nervous, as if aware of the historic importance of the stage on which we was now standing, at the centre. His nervousness eased as he got into his stride.

His presentation justifiably made much of the bewildering complexity of these experiments. Simply to summarise the outcome in terms of a single result – the answer to the Shakespeare question – would not respect the efforts of all involved in running the LHC, operating the detectors, setting the triggers, managing the pile-up of events, calculating the background, managing the worldwide computer grid, performing the detailed analysis, and not sleeping much. Incandela spent quite some time on these technical aspects, as though to reassure everyone that there could be no doubt about the results that he was about to reveal.

When he finally got to it, his punchline was thrilling. Combining the 7 TeV collision data from 2011 and the 8 TeV data from 2012 had produced an excess of events near 125 GeV in the $H \rightarrow \gamma\gamma$ channel with a 4.1-sigma significance. A similar combination of data for the $H \rightarrow Z^0Z^0 \rightarrow \ell^+\ell^-\ell^+\ell^-$ channel had yielded an excess of events with 3.2-sigma significance. Putting the data for these two channels together gave a 5.0-sigma excess. The excess expected of a Standard Model Higgs boson at this mass is 4.7-sigma. 'It's nice to be at five,' Incandela said.[23]

The room erupted in spontaneous applause.

There were some further results to report, relating to other decay channels, but these added little to the overall picture. The combined results are shown in Figure 27(a), in terms of the 'p-value' – a measure of statistical significance of the results – vs. the Higgs mass.

Now rather pressed for time, the seminar swiftly moved on to the second detector collaboration. Fabiola Gianotti stood to present the ATLAS results. She covered much the same ground, emphasising important technical aspects of the experiment. I was struck by one singular fact: with a total of 10.7 inverse femtobarns of data, the number of 126 GeV excess events that could be expected in the $H \rightarrow \gamma\gamma$ channel was estimated to be just 170. The number of background events at this same energy was expected to be 6340, a signal-to-background ratio of just three per cent.

Gianotti's punchline was much the same as her CMS colleague. Combining the 2011 and 2012 data had produced an excess of events at 126.5 GeV in the $H \rightarrow \gamma\gamma$ channel of 4.5-sigma, a significance somewhat larger (by a factor of

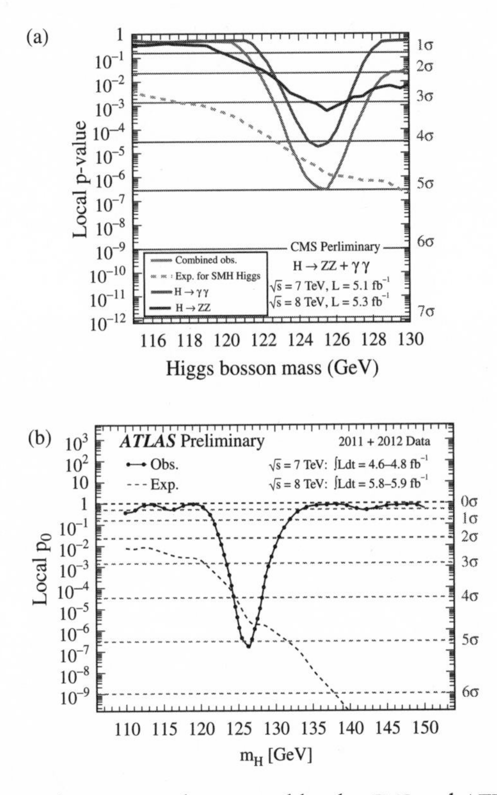


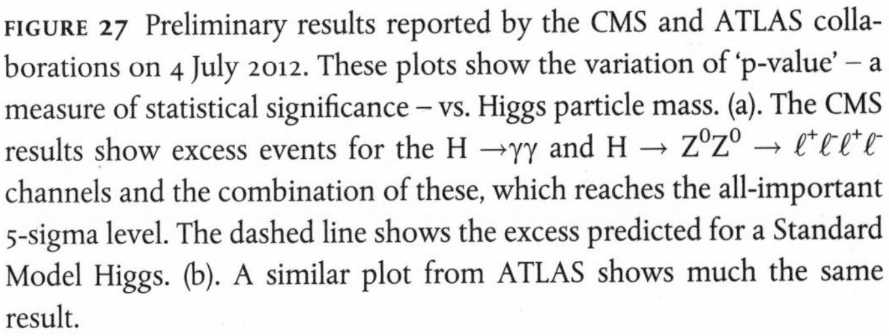

**FIGURE 27** Preliminary results reported by the CMS and ATLAS collaborations on 4 July 2012. These plots show the variation of 'p-value' – a measure of statistical significance – vs. Higgs particle mass. (a). The CMS results show excess events for the H $\rightarrow\gamma\gamma$ and H $\rightarrow$ $Z^0Z^0$ $\rightarrow$ $\ell^+\ell^-\ell^+\ell^-$ channels and the combination of these, which reaches the all-important 5-sigma level. The dashed line shows the excess predicted for a Standard Model Higgs. (b). A similar plot from ATLAS shows much the same result.

two) than the Standard Model prediction. Corresponding data for the $H \rightarrow Z^0Z^0 \rightarrow \ell^+\ell^-\ell^+\ell^-$ channel produced an excess at 125 GeV of 3.4-sigma significance. Combining the data for these two decay channels gave a 5.0-sigma excess, compared to a Standard Model prediction of 4.6-sigma. The results are summarised in Figure 27(b).

Both collaborations had found the 5-sigma evidence sufficient to declare a discovery. More applause.

Heuer declared: 'As a layman I would say that I think we have it. Do you agree?'[24] There could be little doubting that something very much like the Standard Model Higgs boson had been discovered and, to the layman, this was indeed 'it'. But the physicists have more exacting standards. They were now rather cagey about what precisely what kind of discovery they had just announced and, under gentle prodding from journalists in a subsequent press conference, stuck to the conclusion that this new particle was *consistent* with the Higgs. They refused to be drawn on the question of whether or not this indeed was *the* Higgs.

The simple facts are that the new boson has a mass of between 125-126 GeV and interacts with other Standard Model particles in precisely the way expected of the Higgs boson. Apart from the observed enhancement in the $H \rightarrow \gamma\gamma$ decay channel, the new boson's decay modes to other particles have the ratios expected of a Standard Model Higgs. Whilst the ATLAS and CMS experiments are clear that this is a boson, neither is clear on the precise value of its spin quantum number, which could be 0 or 2. However, the only particle anticipated to have spin-2 is the graviton, the purported carrier of the force of gravity. Spin-0 is therefore much more likely. Paraphrasing Rubbia, we might be tempted to declare, with some justification: 'It looks like a Standard Model Higgs, it smells like a Standard Model Higgs, it must be a Standard Model Higgs.'

In truth these results represent a critical milestone on another long journey. A new boson has been discovered that looks to all the world like a Higgs boson. But *which* Higgs boson? The Standard Model needs just one to break the electro-weak symmetry. The Minimal Supersymmetric Standard Model demands five. Other theoretical models make other demands. The only way to find out precisely what kind of particle has been discovered is to explore its properties and behaviour in further experiments.

The CERN press release commented:[25]

> Positive identification of the new particle's characteristics will take considerable time and data. But whatever form the Higgs particle takes, our knowledge of the fundamental structure of matter is about to take a major step forward.

The seminar closed with some thoroughly deserved back-slapping and self-congratulations. When asked for his observations, Peter Higgs congratulated the laboratory on its remarkable success and said: 'It's really an incredible thing that it's happened in my lifetime.'[26]

An important chapter in our efforts to understand the basic nature of material substance is drawing to a close. Another, exciting new chapter is about to begin.

# EPILOGUE
## *The Construction of Mass*

What is the world made of?

In the mid-1930s we would have explained that all the material substance in the world is made of chemical elements and that each element consists of atoms. Each atom consists in turn of a nucleus composed of varying numbers of positively charged protons and electrically neutral neutrons. Surrounding the nucleus are negatively charged electrons, bound by the force of electrical attraction. Each electron can take either a spin-up or spin-down orientation and each atomic orbital can accommodate two electrons provided their spins are paired. Electrons may move from one orbital to another through the absorption or emission of electromagnetic radiation in the form of photons.

We would have explained that the weight of the 18-gram cube of frozen water in the palm of your hand is derived from the collective mass of 10,800 billion trillion protons and neutrons.

Today our answer has become considerably more refined.

The protons and neutrons in the nucleus are not, in fact, elementary particles. They are composed of fractionally charged quarks. A proton consists of three quarks of different 'flavours' – two up and one down. The quarks are also distinguished by their 'colour': red, green, and blue. The two up-quarks and the down-quark in a proton all have different colours, the resulting combination appearing 'white'. A neutron consists of one up-quark and two down-quarks, again with each quark taking a different colour.

The colour force between quarks is carried by eight different kinds of force particle collectively called gluons. This force increases in strength not as the quarks come closer together, as might be expected, but as they separate. The strong nuclear force between protons and neutrons is merely a remnant, a 'hang-over' of the colour force between their constituent quarks.

The discovery of a new particle at CERN suggests strongly that the quark masses are derived from interactions with the Higgs field. These interactions transform otherwise massless quarks into particles with mass. The interactions give the particles *depth*, causing them to slow down. This resistance to acceleration is what we call mass.

But the masses of the quarks are quite small, accounting for just one per cent of the mass of a proton or neutron. The other 99 per cent is derived from the energy carried by the massless gluons which flit between the quarks and bind them together.

In the Standard Model the concept of mass, as an intrinsic property or measure of an amount of substance, has gone. Mass is instead constructed entirely from the *energy* of the interactions that occur between elementary quantum fields and their particles.

The Higgs boson is part of the mechanism that explains how all the mass of all the particles in the universe is constructed. All the matter in the world might consist of quarks and leptons, but it owes its very substance to the energy gained through interactions with the Higgs field and the exchange of gluons.

Without these interactions, matter would be as ephemeral and insubstantial as light itself, and nothing would be.

# ENDNOTES

## Prologue: Form and Substance

1 Albert Einstein, *Annalen der Physik*. **18** (1905), p. 639. English translation quoted in John Stachel (ed.), *Einstein's Miraculous Year: Five Papers that Changed the Face of Physics*, Princeton University Press (2005), p. 161.

## Chapter 1: The Poetry of Logical Ideas

1 Auguste Dick, *Emmy Noether 1882–1935*, Birkhäuser, Boston (1981), p. 32. English translation by H.I. Blocher.
2 Albert Einstein, letter to Hermann Weyl, 8 April 1918, quoted in Pais, *Subtle is the Lord*, p. 341.
3 Louis de Broglie, 'Recherches sur la Théorie des Quanta', PhD Thesis, Faculty of Science, Paris University (1924), p. 10. English translation by A.F. Kracklauer.
4 Albert Einstein, *New York Times*, 5 May 1935.

## Chapter 2: Not a Sufficient Excuse

1 Julian Schwinger, interview with Robert Crease and Charles Mann, 4 March 1983. Quoted in Crease and Mann, p. 127.
2 Richard Feynman, interview with Robert Crease and Charles Mann, 22 February 1985. Quoted in Crease and Mann, p. 139.
3 Freeman Dyson, letter to his parents, 18 September 1948. Quoted in Schweber, p. 505.
4 Feynman, p. 7.

5 Chen Ning Yang, *Selected Papers with Commentary*, W.H. Freeman, New York (1983). Quoted by Christine Sutton in Farmelo (ed.), *It Must be Beautiful*, p. 241.
6 Robert Mills, telephone interview with Robert Crease and Charles Mann, 7 April 1983. Quoted in Crease and Mann, p. 193.
7 Part of a conversation reported by Yang at the International Symposium on the History of Particle Physics, Batavia, Illinois, 2 May 1985. Quoted in Riordan, p. 198.
8 Quoted in Enz, p. 481.
9 Chen Ning Yang, *Selected Papers with Commentary*, W.H. Freeman, New York (1983). Quoted by Christine Sutton in Farmelo (ed.), *It Must be Beautiful*, p. 243.
10 C.N. Yang and R.L. Mills, *Physical Review*, **96**, 1 (1954), p. 195.

## Chapter 3: People Will Be Very Stupid About It

1 Emilio Segrè, *Enrico Fermi: Physicist*, University of Chicago Press (1970), p. 72.
2 Isidor Rabi, quoted in Helge Kragh, *Quantum Generations*, p. 204.
3 Willis Lamb, *Nobel Lectures, Physics 1942–1962*, Elsevier, Amsterdam (1964), p. 286.
4 Quoted by Helge Kragh as 'physics folklore' in *Quantum Generations*, p. 321.
5 Murray Gell-Mann and Edward Rosenbaum, *Scientific American*, July 1957, pp. 72–88. The idea of 'strangeness' was also elaborated around the same time by Japanese physicists Kazuhiko Nishijima and Tadao Nakano (who called it η-charge). Although the term strangeness was retained, the theory is sometimes referred to as Gell-Mann–Nishijima theory.
6 Sheldon Glashow, Harvard University PhD thesis (1958), p. 75. Quoted in Glashow, *Nobel Lectures, Physics 1971–1980*, Edited by Stig Lundqvist, World Scientific, Singapore (1992), p. 496.
7 Murray Gell-Mann, interview with Robert Crease and Charles Mann, 3 March 1983. Quoted in Crease and Mann, p. 225.

8 Murray Gell-Mann, Caltech Report CALT-68-1214, pp. 22–23. Quoted in Crease and Mann, pp. 264–265.

## Chapter 4: Applying the Right Ideas to the Wrong Problem

1 Nambu, p. 180.
2 Robert Serber, telephone interview with Robert Crease and Charles Mann, 4 June 1983. Quoted in Crease and Mann, p. 281.
3 Murray Gell-Mann, interview with Robert Crease and Charles Mann, 3 March 1983. Quoted in Crease and Mann, p. 281.
4 Murray Gell-Mann, interview with Robert Crease and Charles Mann, 3 March 1983. Quoted in Crease and Mann, p. 282.
5 George Zweig, 'An SU(3) Model for Strong Interaction Symmetry and its Breaking', CERN Preprint 8419/TH.412, 21 February 1964, p. 42.
6 P.W. Anderson, *Physical Review*, **130** (1963), p. 441, reproduced in E. Farhi and R. Jackiw (eds.), *Dynamical Gauge Symmetry Breaking: A Collection of Reprints*, World Scientific, Singapore (1982), p. 50.
7 Peter Higgs, in Hoddeson, *et al.*, p. 508.
8 Peter Higgs, *Physical Review Letters*, **13**, 509 (1964).
9 Sidney Coleman, quoted by Peter Higgs in 'My Life as a Boson: the Story of the "Higgs"', presented at the Inaugural Conference of the Michigan Center for Theoretical Physics, 21–25 May 2001.
10 Peter Higgs, in Hoddeson, *et al.*, p. 510.
11 Steven Weinberg, *Nobel Lectures, Physics 1971–1980*, edited by Stig Lundqvist, World Scientific, Singapore (1992), p. 548.
12 Steven Weinberg, interview with Robert Crease and Charles Mann, 7 May 1985. Quoted in Crease and Mann, p. 245.

## Chapter 5: I Can Do That

1 Steven Weinberg, quoted by John Iliopoulos in an interview with Michael Riordan, 4 June 1985. Quoted in Riordan, p. 211.

2 Sheldon Glashow, *Nobel Lectures, Physics 1971–1980*, edited by Stig Lundqvist, World Scientific, Singapore (1992), p. 500.
3 Gerard 't Hooft, *In Search of the Ultimate Building Blocks*, Cambridge University Press (1997), p. 58.
4 Martinus Veltman, private communication to Andrew Pickering, quoted in Pickering, p. 178.
5 Gerard 't Hooft, interview with Robert Crease and Charles Mann, 26 September 1984. Quoted in Crease and Mann, pp. 325–6.
6 Martinus Veltman in Hoddeson, *et al.*, p. 173.
7 Sheldon Glashow, quoted by David Politzer, interview with Robert Crease and Charles Mann, 21 February 1985. Quoted in Crease and Mann, p. 326.
8 Gerard 't Hooft, in Hoddeson, *et al.*, p. 192.
9 Murray Gell-Mann, in Hoddeson, *et al.*, p. 629.
10 W.A. Bardeen, H. Fritzsch, and M. Gell-Mann, *Proceedings of the Topical Meeting on Conformal Invariance in Hadron Physics*, Frascati, May 1972. Quoted in Crease and Mann, p. 328.
11 Murray Gell-Mann in Hoddeson, *et al.*, p. 631.

## Chapter 6: Alternating Neutral Currents

1 Richard Feynman, interview with Michael Riordan, 14–15 March 1984. Quoted in Riordan, p. 152.
2 Richard Feynman, interview with Paul Tsai, 3 April 1984. Quoted in Riordan, p. 150.
3 Richard Feynman, quoted by Jerome Friedman in an interview with Michael Riordan, 24 October 1985. Quoted in Riordan, p. 151.
4 Donald Perkins in Hoddesson *et al.*, p. 430.
5 Carlo Rubbia, letter to Andre Lagarrigue, 17 July 1973. Quoted in Crease and Mann, p. 352.
6 Donald Perkins, *CERN Courier*, 1 June 2003.
7 David Cline, quoted in Crease and Mann, p. 357.

## Chapter 7: They Must Be Ws

1 W.A. Bardeen, H. Fritzsch, and M. Gell-Mann, *Proceedings of the Topical Meeting on Conformal Invariance in Hadron Physics*, Frascati, May 1972. Quoted in Crease and Mann, p. 328.
2 Frank Wilczek, *MIT Physics Annual 2003*, p. 35.
3 Pierre Darriulat, in Cashmore *et al.*, p. 57.
4 Simon van der Meer, quoted in Brian Southworth and Gordon Fraser, *CERN Courier*, November 1983.
5 Pierre Darriulat, in Cashmore *et al.*, p. 57.
6 Carlo Rubbia, quoted in Brian Southworth and Gordon Fraser, *CERN Courier*, November 1983.
7 Lederman, p. 357.

## Chapter 8: Throw Deep

1 Howard Georgi and Sheldon Glashow, *Physical Review Letters*, **32** (1974), p. 438.
2 Howard Georgi, interview with Robert Crease and Charles Mann, 29 January 1985. Quoted in Crease and Mann, p. 400.
3 Guth, p. 176.
4 *New York Times*, 6 June 1983.
5 The full quotation reads: I would rather be ashes than dust; I would rather my spark should burn out in a brilliant blaze; Than it should be stifled in dry rot; I would rather be a superb meteor; With every atom of me in magnificent glow; Than a sleepy and permanent planet. Jack London, quoted in Halpern, p. 151.
6 Attributed to (or associated with) Ken Stabler. The quote was used by journalist George Will in the title of an article about Reagan's support for the SSC which subsequently appeared in the *Washington Post*.
7 This short speech from the 1940 film *Knute Rockne: All American* can be found on the American Rhetoric website at: www.americanrhetoric.com/MovieSpeeches/moviespeechknuterockneallamerican.html

8 Weinberg, p. 220.
9 Lederman, p. 406.
10 Raphael Kasper, quoted in *Dallas Morning News*, 23 July 2005.
11 Herman Wouk, *A Hole in Texas*, Little, Brown & Company, New York (2004), Author's Note.
12 Carlo Rubbia, quoted in Lederman, p. 381.

## Chapter 9: A Fantastic Moment

1 William Waldegrave, quoted in Sample, p. 163.
2 David Miller's submission is available at: http://www.hep.ucl.ac.uk/~djm/higgsa.html. Quoted with permission.
3 David Miller, personal communication, 4 October 2010.
4 Luciano Maiani, *CERN Courier*, 26 February 2001.
5 http://cms.web.cern.ch/cms/Detector/FullDetector/index.html
6 Lyndon Evans, quoted in *CERN Bulletin* 37–38, 2008.

## Chapter 10: The Shakespeare Question

1 *Fermilab Today* Twitter feed, quoted by Tom Chivers, *The Telegraph*, 13 July 2010.
2 Tommaso Dorigo, 'Rumours About a Light Higgs', *A Quantum Diaries Survivor*, blog entry 8 July 2010, www.science20.com/quantum_diaries_survivor/
3 Leon Lederman, quoted by Tom Chivers, *The Telegraph*, 13 July 2010.
4 Rolf Heuer, quoted in *CERN Bulletin*, Monday 31 January 2011.
5 Albert Einstein, quoted in Alice Calaprice (ed.), *The Ultimate Quotable Einstein*, Princeton University Press,.2011, p. 409.
6 Jon Butterworth, television interview with Krishnan Guru-Murthy, *Channel 4 News*, 24 April 2011.
7 Jon Butterworth, 'Rumours of the Higgs at ATLAS', *Life and Physics*, hosted by the *Guardian*, blog entry 24 April 2011. www.guardian.co.uk/science/life-and-physics

8 David Shiga, 'Elusive Higgs Slips from Sight Again', *New Scientist*, 4 May 2011.
9 Jon Butterworth, 'Told You So… Higgs Fails to Materialise', *Life and Physics*, hosted by the *Guardian*, blog entry 11 May 2011. www.guardian.co.uk/science/life-and-physics
10 Laurette Ponce, interview with the author, 21 June 2011.
11 Rolf Heuer, *DG's Talk to Staff*, CERN, 4 July 2011.
12 Lyndon Evans, interview with the author, 22 June 2011.
13 Rolf Heuer, *DG's Talk to Staff*, CERN, 4 July 2011.
14 Peter Higgs, interview with the author, 18 August 2011.
15 CERN Press Release, 22 August 2011.
16 Fabiola Gianotti, quoted in CERN Press Release, 13 December 2011.
17 Rolf Heuer, closing remarks, CERN Public Seminar, 13 December 2011.
18 Jon Butterworth, television interview with Jon Snow, *Channel 4 News*, 13 December 2011.
19 Peter Higgs, quoted by Alan Walker, communication to the author, 13 December 2011.
20 Tommaso Dorigo, 'Firm Evidence of a Higgs Boson at Last!', *A Quantum Diaries Survivor*, blog entry 13 December 2011, www.science20.com/quantum_diaries_survivor/
21 Matt Strassler, 'Higgs Update Today: Inconclusive, as Expected', Of Particular Significance, comment on blog entry 13 December 2011, profmattstrassler.com/2011/12/13/
22 Jon Butterworth, communication to the author, 23 December 2011.
23 Joe Incandela, 'Latest update in the search for the Higgs boson', CERN Seminar, 4 July 2012.
24 Rolf Heuer, 'Latest update in the search for the Higgs boson', CERN Seminar, 4 July 2012.
25 CERN Press Release, 4 July 2012.
26 Peter Higgs, 'Latest update in the search for the Higgs boson', CERN Seminar, 4 July 2012.

# GLOSSARY

**Anti-particle**. Identical in mass to an 'ordinary' particle but of opposite charge. For example, the anti-particle of the electron ($e^-$) is the positron ($e^+$). The anti-particle of a red quark is an anti-red anti-quark. Every particle in the Standard Model has an anti-particle. Particles with zero charge are their own anti-particles.

**Asymptotic freedom**. A property of the strong colour force between quarks. The colour force actually declines in strength as quarks are brought closer together, such that in the asymptotic limit of zero separation the quarks behave as though they are completely free – see Figure 17(b), p. 136.

**ATLAS**. Acronym for A Toroidal LHC Apparatus, one of the two detector collaborations involved in the hunt for the Higgs boson at CERN's Large Hadron Collider.

**Atom**. From the Greek *atomos*, meaning indivisible. Originally intended to denote the ultimate constituents of matter, the word atom now signifies the fundamental constituents of individual chemical elements. Thus, water consists of molecules of $H_2O$, which is composed of two atoms of hydrogen and one atom of oxygen. The atoms in turn consist of protons and neutrons, which are bound together to form a central nucleus, and electrons whose wavefunctions form characteristic patterns called orbitals around the nucleus.

**Baryon**. From the Greek *barys*, meaning heavy. Baryons form a subclass of hadrons. They are heavier particles which experience the strong nuclear force and include the proton and neutron. They are composed of triplets of quarks.

**Beta-particle**. A high-speed electron emitted from the nucleus of an atom undergoing beta-radioactive decay. *See* beta-radioactivity/decay.

**Beta-radioactivity/decay**. First discovered by French physicist Henri Becquerel in 1896 and so named by Ernest Rutherford in 1899. An example of a weak-force decay, it involves transformation of a down-quark in a neutron into an up-quark, turning the neutron into a proton with the emission of a $W^-$ particle. The $W^-$ decays into a high-speed electron (the 'beta-particle') and an electron anti-neutrino.

**Big bang**. Term used to describe the cosmic 'explosion' of space-time and matter during the early moments in the creation of the universe, about 13.7 billion years ago. Originally coined by maverick physicist Fred Hoyle as a derogatory term, overwhelming evidence for a big bang 'origin' of the universe has since been obtained through the detection and mapping of the cosmic microwave background radiation, the cold remnant of hot radiation thought to have disengaged from matter about 380,000 years after the big bang.

**Billion**. One thousand million, $10^9$, or 1,000,000,000.

**Boson**. Named for Indian physicist Satyendra Nath Bose. Bosons are characterized by integral spin quantum numbers (1, 2, . . . , etc.) and, as such, are not subject to Pauli's exclusion principle. Bosons are involved in the transmission of forces between matter particles, and include the photon (electromagnetism), the W and Z particles (weak force), and gluons (colour force). Particles with spin zero are also called bosons but these are not involved in transmitting forces. Examples include the pions, Cooper pairs (which can also have spin 1), and the Higgs boson. The graviton, the hypothetical particle of the gravitational field, is believed to be a boson with spin 2.

**Bottom quark**. Also sometimes referred to as the 'beauty' quark. A third-generation quark with charge $-\frac{1}{3}$, spin ½ (fermion), and a 'bare mass' of 4.19 GeV. It was discovered at Fermilab in 1977, through the observation of the upsilon, a meson formed from bottom and anti-bottom quarks.

**CERN**. Acronym for Conseil Européen pour la Recherche Nucléaire (the European Council for Nuclear Research), founded in 1954. This was renamed the Organisation Européenne pour la Recherche Nucléaire (European Organization for Nuclear Research) when the provisional Council was dissolved, but the acronym CERN was retained. CERN is located in the north-west suburbs of Geneva near the Swiss–French border.

**Charm-quark**. A second-generation quark with charge $+\frac{2}{3}$, spin ½ (fermion), and a 'bare mass' of 1.27 GeV. It was discovered simultaneously at Brookhaven National Laboratory and SLAC in the 'November revolution' of 1974 through the observation of the $J/\psi$, a meson formed from a charm- and an anti-charm-quark.

**CMS**. Acronym for Compact Muon Solenoid, one of the two detector collaborations involved in the hunt for the Higgs boson at CERN's Large Hadron Collider.

**Cold dark matter** (CDM). A key component of the current lambda-CDM model of big bang cosmology, thought to account for about 22 per cent of the mass-energy of the universe. The constitution of cold dark matter is unknown, but is thought to consist largely of 'non-baryonic' matter, i.e. matter that does not involve protons or neutrons, most likely particles not currently known to the Standard Model. Candidates include weakly interacting massive particles, or WIMPs. They have many of the properties of neutrinos, but are required to be far more massive and therefore move much more slowly. Supersymmetric extensions of the Standard Model suggest that such particles might be neutralinos.

**Colour charge**. A property possessed by quarks in addition to flavour (up, down, strange, etc.). Unlike electric charge, which comes in two varieties – positive and negative – colour charge comes in three varieties – red, green, and blue. Obviously, the use of these names does not imply that quarks are 'coloured' in the conventional sense. The colour force between quarks is carried by coloured gluons.

**Colour force**. The strong force responsible for binding quarks and gluons together inside hadrons. Unlike more familiar forces, such as gravity and electromagnetism, the colour force exhibits asymptotic freedom – at the asymptotic limit of zero separation, quarks behave as though they are entirely free. The strong nuclear force which binds protons and neutrons together inside atomic nuclei is thought to be a 'hang-over' of the colour force binding quarks inside the nucleons.

**Complex number**. A complex number is formed by multiplying a real number by the square-root of –1, written *i*. The square of a complex number is then a negative number for example, the square of 5*i* is –25. Complex numbers are used widely in mathematics to solve problems that are impossible using real numbers only.

**Conservation law**. A physical law which states that a specific measureable property of an isolated system does not change as the system evolves in time. Measureable properties for which conservation laws have been established include mass-energy, linear and angular momentum, electric and colour charge, isospin, etc. According to Noether's theorem, each conservation law can be traced to a specific continuous symmetry of the system.

**Cooper pair**. When cooled below its critical temperature, electrons in a superconductor experience a weak mutual attraction. Electrons with opposite spin and momentum combine to form Cooper pairs, which move through the metal lattice cooperatively, their motion mediated or facilitated by lattice vibrations. Such pairs of electrons have spin 0 or 1 and are therefore bosons. Consequently, there is no restriction on the number of pairs that can occupy a single quantum state, and at low temperatures they can 'condense', building the state to macroscopic dimensions. The Cooper pairs in this state experience no resistance as they pass through the lattice and the result is superconductivity.

**Cosmic inflation**. A rapid exponential expansion of the universe thought to have occurred between $10^{-36}$ and $10^{-32}$ seconds after the big bang. Discovered in the context of GUTs by American physicist Alan Guth in 1980, inflation helps to explain the large-scale structure of the universe that we observe today.

**Cosmic microwave background radiation**. Some 380,000 years after the big bang, the universe had expanded and cooled sufficiently to allow hydrogen nuclei (protons) and helium nuclei (consisting of two protons and two neutrons) to recombine with electrons to form neutral hydrogen and helium atoms. At this point, the universe became 'transparent' to the residual hot radiation. Further expansion has shifted and cooled this hot radiation to the microwave region with a temperature of just 2.7 K (−270.5 °C), a few degrees above absolute zero. This microwave background radiation was predicted by several theorists and was discovered accidentally by Arno Penzias and Robert Wilson in 1964. The COBE and WMAP satellites have since studied this radiation in detail.

**Cosmic rays**. Streams of high-energy charged particles from outer space which wash constantly over the earth's upper atmosphere. The use of the term 'ray' harks back to the early days of research on radioactivity, when directed streams of charged particles were referred to as 'rays'. Cosmic rays are derived from a variety of sources, including high-energy processes occurring on the surface of the sun and other stars, and as-yet unknown processes occurring elsewhere in the universe. The energies of cosmic ray particles are typically between 10 MeV and 10 GeV.

**Cosmological constant**. In 1922 Russian theorist Alexander Friedmann found solutions to Einstein's gravitational field equations that describe a universe in which space-time is expanding. Einstein had initially resisted the idea that space-time could expand or contract and had fudged his equations to produce static solutions. Concerned that conventional gravity would be expected to overwhelm the matter in the universe and cause it to collapse in on itself, Einstein had introduced a 'cosmological constant' – a kind of negative or repulsive form of gravity – to counteract the effect. When evidence accumulated that the universe is actually expanding, Einstein regretted his action, calling it the biggest blunder he had ever made in his life. But, in fact, further discoveries in 1998 suggested that the expansion of the universe is actually accelerating. When combined with satellite measurements of the cosmic microwave background radiation these results have led to the suggestion that the universe is pervaded by 'dark energy', accounting for about 73 per cent of the mass-energy of the universe. One form of dark energy requires the reintroduction of Einstein's cosmological constant.

**Dark matter**. Discovered in 1934 by Swiss astronomer Fritz Zwicky as an anomaly in the measured masses of galaxies in the Coma Cluster (located in the constellation Coma Berenices) based on the observed motions of the galaxies near the cluster edge compared with the number of observable galaxies and the total brightness of the cluster. These estimates of the masses of the galaxies differed by a factor of 400. As much as 90 per cent of the mass required to explain the size of the gravitational effects appeared to be 'missing', or invisible. This missing matter was called 'dark matter'. Subsequent studies favour a form of dark matter called 'cold dark matter'. *See* cold dark matter.

**Deep inelastic scattering**. A kind of particle scattering event in which much of the energy of the accelerated particle (for example, an electron) is channelled into the destruction of the target particle (for example, a proton). The accelerated particle recoils from the collision with considerably less energy and a spray of different hadrons is produced.

**Degree of freedom**. The number of dimensions that are accessible to a system or in which the system is free to move. A classical particle is free to move in three spatial dimensions. However, photons are massless particles with spin 1 and, as such, are constrained to only two dimensions, manifested as left and right circular polarization or vertical and horizontal polarization. In the Higgs mechanism, massless bosons may gain a third degree of freedom by absorbing a Nambu–Goldstone boson, see Figure 14, p. 89.

**Eightfold Way**. A scheme for classifying the 'zoo' of particles known around 1960 in the form of two 'octets', developed by Murray Gell-Mann and independently by Yuval Ne'eman. The patterns are based on a global SU(3) symmetry and are formed by mapping the particles according to their electric charge or total isospin vs. strangeness (see Figure 10, p. 69). The patterns were eventually explained by the quark model (Figure 12, p. 82).

**Electric charge**. A property possessed by quarks and leptons (and, more familiarly, protons and electrons). Electric charge comes in two varieties – positive and negative – and the flow of negative charge is the basis for electricity and the power industry.

**Electromagnetic force**. Electricity and magnetism were recognized to be components of a single, fundamental force, through the work of several experimental and theoretical physicists, most notably English physicist Michael Faraday and Scottish theoretician James Clerk Maxwell. The electromagnetic force is responsible for binding electrons with their nuclei inside atoms, and binding atoms together to form the great variety of molecular substances.

**Electron**. Discovered in 1897 by English physicist J.J. Thompson. The electron is a first-generation lepton with a charge –1, spin ½ (fermion), and mass 0.51 MeV.

**Electron volt (eV)**. An electron volt is the amount of energy a single negatively charged electron gains when accelerated through a one-volt electric field. A 100 W light bulb burns energy at the rate of about 600 billion billion electron volts per second.

**Electro-weak force**. Despite the great difference in scale between the electromagnetic and weak nuclear forces, these are facets of what was once a unified electro-weak force, thought to prevail during the 'electro-weak epoch', between $10^{-36}$ and $10^{-12}$ seconds after the big bang. The combination of electromagnetic and weak nuclear forces in an $SU(2) \times U(1)$ field theory was first achieved by Steven Weinberg and independently by Abdus Salam in 1967–68.

**Element**. The philosophers of Ancient Greece believed that all material substance is composed of four elements – earth, air, fire, and water. A fifth element, variously called the ether or 'quintessence', was introduced by Aristotle to describe the unchanging heavens. Today, these classical elements have been replaced by a system of chemical elements. These are 'fundamental' in the sense that chemical elements cannot be transformed one into another by chemical means, meaning that they consist of only one type of atom. The elements are organized in a 'periodic table', from hydrogen to uranium and beyond.

**Exclusion principle**. *See* Pauli exclusion principle.

**Fermion**. Named for Italian physicist Enrico Fermi. Fermions are characterized by half-integral spins ($\frac{1}{2}$, $\frac{3}{2}$, etc.) and include quarks and leptons and many composite particles produced from various combinations of quarks, such as baryons.

**Flavour**. A property which distinguishes one type of quark from another, in addition to colour charge. There are six flavours of quark which form three generations up, charm, and top with electric charge $+\frac{2}{3}$, spin ½, and masses of 1.7–3.3 MeV, 1.27 GeV, and 172 GeV, respectively, and down, strange, and bottom with electric charge $-\frac{1}{3}$, spin ½, and masses 4.1–5.8 MeV, 101 MeV, and 4.19 GeV, respectively. The term flavour is also applied to leptons, with the electron, muon, tau, and their corresponding neutrinos distinguished by their 'lepton flavour'. *See* lepton.

**Gauge symmetry**. A name coined by German mathematician Hermann Weyl. When applied to quantum field theories, a 'gauge' is chosen to which the equations are invariant – arbitrary changes in the gauge make no difference to the predicted outcomes. The link between gauge symmetry and conservation laws (*see* conservation law and Noether's theorem) means that the correct choice of gauge symmetry can lead to a field theory which will automatically respect the need for conservation of the property under study.

**Gauge theory**. A gauge theory is one based on a gauge symmetry (*see* gauge symmetry). Einstein's general theory of relativity is a gauge theory which is invariant to arbitrary changes in the space-time coordinate system (the 'gauge'). Quantum electrodynamics (QED) is a quantum field theory which is invariant to the phase of the electron wavefunction. In the 1950s developing quantum field theories of the strong and weak nuclear forces became a matter of identifying the conserved quantity and hence the appropriate gauge symmetry.

**General relativity**. Developed by Einstein in 1915, the general theory of relativity incorporates special relativity and Newton's law of universal gravitation in a geometric theory of gravitation. Einstein replaced the 'action-at-a-distance' implied in Newton's theory of universal gravitation with the movement of massive bodies in a curved space-time. In general relativity, matter tells space-time how to curve, and the curved space-time tells matter how to move.

**g-factor**. A constant of proportionality between the (quantized) angular momentum of an elementary or composite particle and its magnetic moment, the direction the particle will adopt in a magnetic field. There are actually three *g*-factors for the electron: one associated with its spin, one associated with the angular momentum of the electron orbital motion in an atom, and one associated with the sum of spin and orbital angular momentum. Dirac's relativistic quantum theory of the electron predicted a *g*-factor for electron spin of 2. The value recommended by the CODATA task group in 2006 is 2.0023193043622. The difference is due to quantum electrodynamic effects.

**Giga**. A prefix denoting billion. A giga electron volt (GeV) is a billion electron volts, $10^9$ eV, or 1000 MeV.

**Gluon**. The carrier of the strong colour force between quarks. Quantum chromodynamics requires eight, massless colour force gluons which themselves carry colour charge. Consequently, the gluons participate in the force rather than simply transmit it from one particle to another. Ninety-nine per cent of the mass of protons and neutrons is thought to be energy carried by gluons.

**Grand unified theory (GUT)**. Any theory which attempts to unify the electromagnetic, weak, and strong nuclear forces in a single structure is an example of a grand unified theory. The first example of a GUT was developed by Sheldon Glashow and Howard Georgi in 1974. GUTs do not seek to accommodate gravity; theories that do are generally referred to as Theories of Everything (TOEs).

**Gravitational force**. The force of attraction experienced between all mass-energy. Gravity is extremely weak, and has no part to play in the interactions between atoms, sub-atomic, and elementary particles which are rather governed by the colour force, weak nuclear force, and electromagnetism. The force of gravity is described by Einstein's general theory of relativity.

**Graviton**. A hypothetical particle which carries the gravitational force in quantum field theories of gravity. Although many attempts have been made to develop such a theory, to date these have not been recognized as successful. If it exists, the graviton would be a massless, chargeless boson with spin 2.

**Hadron**. From the Greek *hadros*, meaning thick or heavy. Hadrons form a class of particles which experience the strong nuclear force and are therefore composed of various combinations of quarks. This class includes baryons, which are composed of three quarks, and mesons, which are composed of one quark and an anti-quark.

**Higgs boson**. Named for English physicist Peter Higgs. All Higgs fields have characteristic field particles called Higgs bosons. The term 'Higgs boson' is typically reserved for the electro-weak Higgs, the particle of the Higgs field first used in 1967–68 by Steven Weinberg and Abdus Salam to account for electro-weak symmetry-breaking. Something that looks very much like the electro-weak Higgs boson was discovered at CERN's Large Hadron Collider on 4 July 2012. It is a neutral, spin 0 particle with a mass of 125 GeV.

**Higgs field**. Named for English physicist Peter Higgs. A generic term used for any background energy field added to a quantum field theory to trigger symmetry-breaking through the Higgs mechanism. The existence of the Higgs field used to break the symmetry in a quantum field theory of the electro-weak force is strongly supported by the discovery of the new particle at CERN.

**Higgs mechanism**. Named for English physicist Peter Higgs, but also often referred to using the names of other physicists who independently discovered the mechanism in 1964. One alternative name is the Brout–Englert–Higgs–Hagen–Guralnik–Kibble – BEHHGK, or 'beck' mechanism, after the physicists Robert Brout, François Englert, Peter Higgs, Gerald Guralnik, Carl Hagen, and Tom Kibble. The mechanism describes how a background field – called the Higgs field – can be added to a quantum field theory to break a symmetry of the theory. In 1967–68 Steven Weinberg and Abdus Salam independently used the mechanism to develop a field theory of the electro-weak force.

**Inflation**. *See* cosmic inflation.

**Isospin**. Also known as isotopic or isobaric spin. Introduced by Werner Heisenberg in 1932 to explain the symmetry between the newly discovered neutron and the proton. Isospin symmetry is now understood to be a subset of the more general flavour symmetry in hadron interactions. The isospin of a particle can be calculated from the number of up- and down-quarks it contains (see p. 81).

**Kaon**. A group of spin-0 mesons consisting of up-, down-, and strange-quarks and their anti-quarks. These are $K^+$ (up-anti-strange), $K^-$ (strange-anti-up), and $K^0$ (mixtures of down-anti-strange and strange-anti-down) with masses 494 MeV ($K^{\pm}$) and 498 MeV ($K^0$).

**Lambda-CDM**. An abbreviation of lambda-cold dark matter. Also known as the 'Standard Model' of big bang cosmology. The lambda-CDM model accounts for the large-scale structure of the universe, the cosmic microwave background radiation, the accelerating expansion of the universe, and the distribution of elements such as hydrogen, helium, lithium, and oxygen. The model assumes that 73 per cent of the mass-energy of the universe is dark energy (reflected in the size of the cosmological constant, lambda), 22 per cent is cold dark matter, leaving the visible universe – galaxies, stars, and known planets – to account for just 5 per cent.

**Lamb shift**. A small difference between two electron energy levels of the hydrogen atom, discovered by Willis Lamb and Robert Retherford in 1947. The Lamb shift provided an important clue which led to the development of renormalization and ultimately quantum electrodynamics.

**LEP**. Acronym for Large Electron–Positron collider, the predecessor of the LHC at CERN.

**Lepton**. From the Greek *leptos*, meaning small. Leptons form a class of particles which do not experience the strong nuclear force and combine with quarks to form matter. Like quarks, leptons form three generations, including the electron, muon, and tau, with electric charge –1, spin ½, and masses 0.51 MeV, 106 MeV, and 1.78 GeV, respectively, and their corresponding neutrinos. The electron, muon, and tau neutrinos carry no electric charge, have spin ½, and are believed to possess very small masses (necessary to explain the phenomenon of neutrino oscillation, the quantum-mechanical mixing of neutrino flavours such that the flavour may change over time).

**LHC**. Acronym for Large Hadron Collider. The world's highest-energy particle accelerator, capable of producing proton–proton collision energies of 14 TeV. The LHC is 27 kilometres in circumference and lies 175 metres beneath the Swiss–French border at CERN near Geneva. The LHC, operating at proton–proton collision energies of 7 TeV and subsequently 8 TeV, provided the evidence which led to the discovery of a new, Higgs-like boson in July 2012.

**Luminosity**. The luminosity of a beam of particles in an accelerator is the number of particles per unit area per unit time multiplied by the opacity of the beam target (a measure of the impenetrability of the target to the particles). Of particular interest is the integrated luminosity, which is simply the integral (sum) of the luminosity over time, usually reported in units of per square centimetre ($cm^{-2}$) or inverse barns ($10^{24}$ $cm^{-2}$). The number of collisions which result in a particular elementary particle reaction is then just the integrated luminosity multiplied by the cross-section (in units of $cm^{2}$) for the reaction, which is a measure of its likelihood.

**Mega**. A prefix denoting million. A mega electron volt (MeV) is a million electron volts, $10^{6}$ eV or 1,000,000 eV.

**Meson**. From the Greek *mésos*, meaning 'middle'. Mesons are a subclass of hadrons. They experience the strong nuclear force and are composed of quarks and anti-quarks.

**MIT**. Acronym for the Massachusetts Institute of Technology.

**Mole**. A standard unit for the amount of a chemical substance, equal to its atomic or molecular weight in grams. A mole contains $6.022 \times 10^{23}$ particles. The name is derived from 'molecule'.

**Molecule**. A fundamental unit of chemical substance formed from two or more atoms. A molecule of oxygen consists of two oxygen atoms, $O_2$. A molecule of water consists of two hydrogen atoms and one oxygen atom, $H_2O$.

**MSSM**. Acronym for the Minimum Supersymmetric Standard Model, the minimal extension of the conventional Standard Model of particle physics which accommodates supersymmetry, developed in 1981 by Howard Georgi and Savas Dimopoulos.

**Muon**. A second-generation lepton equivalent to the electron, with a charge −1, spin ½ (fermion), and mass 106 MeV. First discovered in 1936 by Carl Anderson and Seth Neddermeyer.

**NAL**. Acronym for the National Accelerator Laboratory in Chicago. Renamed the Fermi National Accelerator Laboratory, or 'Fermilab', in 1974.

**Nambu–Goldstone boson**. A massless, spin-0 particle created as a consequence of spontaneous symmetry-breaking, first discovered by Yoichiro Nambu in 1960 and elaborated by Jeffrey Goldstone in 1961. In the Higgs mechanism, the Nambu–Goldstone bosons become a third 'degree of freedom' of quantum particles that would otherwise be massless (see Figure 14, p. 89).

**Neutral currents (weak force)**. Interactions between elementary particles involving no change in electric charge. These may involve exchange of a virtual $Z^0$ particle or simultaneous exchange of both $W^+$ and $W^-$ particles (see Figures 15 and 16, pp. 100 and 127).

**Neutrino**. From Italian, meaning 'small neutral one'. Neutrinos are the chargeless, spin ½ (fermion) companions to the negatively charged electron, muon, and tau. The neutrinos are believed to possess very small masses, necessary to explain the phenomenon of neutrino oscillation, the quantum-mechanical mixing of neutrino flavours such that the flavour may change over time. Neutrino oscillation solves the solar neutrino problem – that the numbers of neutrinos measured to pass through the earth are inconsistent with the numbers of electron neutrinos expected from nuclear reactions occurring in the sun's core. It was determined in 2001 that only 35 per cent of the neutrinos from the sun are electron neutrinos – the balance are muon and tau neutrinos, indicating that the neutrino flavours oscillate as they travel from the sun to the earth.

**Neutron**. An electrically neutral sub-atomic particle, first discovered in 1932 by James Chadwick. The neutron is a baryon consisting of one up- and two down-quarks with spin ½ and mass 940 MeV.

**Noether's theorem**. Developed by Amalie Emmy Noether in 1918, the theorem connects the laws of conservation with specific continuous symmetries of physical systems and the theories that describe them, used as a tool in the development of new theories. The conservation of energy reflects the fact that the laws governing energy are invariant to continuous changes or 'translations' in time. For linear momentum, the laws are invariant to continuous translations in space. For angular momentum, the laws are invariant to the *angle* of direction measured from the centre of the rotation.

**Nucleus**. The dense region at the core of an atom in which most of the atom's mass is concentrated. Atomic nuclei consist of varying numbers of protons and neutrons. The nucleus of a hydrogen atom consists of a single proton.

**Parton**. A name coined by Richard Feynman in 1968 to describe the point-like constituent 'parts' of protons and neutrons. Partons were subsequently shown to be quarks and gluons.

**Pauli exclusion principle**. Discovered by Wolfgang Pauli in 1925. The exclusion principle states that no two fermions may occupy the same quantum state (i.e. possess the same set of quantum numbers) simultaneously. For electrons, this means that only two electrons can occupy a single atomic orbital provided that they possess opposite spins.

**Perturbation theory**. A mathematical method used to find approximate solutions to equations that cannot be solved exactly. The offending equation is recast as a perturbation expansion – the sum of a potentially infinite series of terms which starts with a 'zeroth-order' expression which can be solved exactly. To this are added additional (or perturbation) terms representing corrections to first-order, second-order, third-order, etc. In principle, each term in the expansion provides a smaller and smaller correction to the zeroth-order result, gradually bringing the calculation closer and closer to the actual result. The accuracy of the final result then depends simply on the number of perturbation terms included in the calculation. Although it is structurally very different, we can get some idea of how the perturbation expansion is supposed to work by looking at the power series expansion for a simple trigonometric function such as sin $x$. The first few terms in the expansion are: $\sin x = x - x^3/3! + x^5/5! - x^7/7! + \cdots$ For $x = 45°$ (0.785398 radians), the first term gives 0.785398 from which we subtract 0.080745, then add 0.002490, then subtract 0.000037. Each successive term gives a smaller correction, and after just four terms we have the result 0.707106, which should be compared with $\sin(45°) = 0.707107$.

**Photon**. The elementary particle underlying all forms of electromagnetic radiation, including light. The photon is a massless, spin 1 boson which acts as the carrier of the electromagnetic force.

**Pion**. A group of spin-0 mesons formed from up- and down-quarks and their anti-quarks. These are $\pi^+$ (up-anti-down), $\pi^-$ (down-anti-up), and $\pi^0$ (a mixture of up-anti-up and down-anti-down), with masses 140 MeV ($\pi^\pm$) and 135 MeV ($\pi^0$).

**Planck constant**. Denoted $h$. Discovered by Max Planck in 1900. The Planck constant is a fundamental physical constant which reflects the magnitudes of quanta in quantum theory. For example, the energies of photons are determined by their radiation frequencies according to the relation $E = h\nu$, i.e. energy equals Planck's constant multiplied by the radiation frequency. Planck's constant has the value $6.626 \times 10^{-34}$ joule seconds.

**Positron**. The anti-particle of the electron, denoted $e^+$, with a charge +1, spin ½ (fermion), and mass 0.51 MeV. The positron was the first anti-particle to be discovered, by Carl Anderson in 1932.

**Proton**. A positively charged sub-atomic particle 'discovered' and so named by Ernest Rutherford in 1919. Rutherford actually identified that the nucleus of the hydrogen atom (which is a single proton) is a fundamental constituent of other atomic nuclei. The proton is a baryon consisting of two up- and one down-quarks with spin ½ and mass 938 MeV.

**Quantum**. A fundamental, indivisible unit of properties such as energy and angular momentum. In quantum theory, such properties are recognized not to be continuously variable but to be organized in discrete packets or bundles, called quanta. The use of the term is extended to include particles. Thus, the photon is the quantum particle of the electromagnetic field. This idea can be extended beyond the carriers of forces to include matter particles themselves. Thus, the electron is the quantum of the electron field, and so on. This is sometimes referred to as second quantization.

**Quantum chromodynamics (QCD)**. The SU(3) quantum field theory of the strong colour force between quarks carried by a system of eight coloured gluons.

**Quantum electrodynamics (QED)**. The U(1) quantum field theory of the electromagnetic force between electrically charged particles, carried by photons.

**Quantum field**. In classical field theory a 'force field' is ascribed a value at every point in space-time and can be scalar (magnitude but no direction) or vector (magnitude and direction). The 'lines of force' made visible by sprinkling iron filings on a piece of paper held above a bar magnetic provides a visual representation of such a field. In a quantum field theory, forces are conveyed by ripples in the field which form waves and – because waves can also be interpreted as particles – as quantum particles of the field. This idea can be extended beyond the carriers of forces (bosons) to include matter particles (fermions). Thus, the electron is the quantum of the electron field, and so on.

**Quantum number**. The description of the physical state of a quantum system requires the specification of its properties in terms of total energy, linear and angular momentum, electric charge, etc. One consequence of the quantization of such properties is the appearance in this description of regular multiples of the associated quanta. For example, the angular momentum associated with the spin of an electron is fixed at the value $\frac{1}{2} h/2\pi$, where $h$ is Planck's constant. The recurring integral or half-integral numbers which multiply the sizes of the quanta are called quantum numbers. When placed in a magnetic field, the electron spin may be oriented along or against the field lines of force, giving rise to 'spin-up' and 'spin-down' orientations characterized by the quantum numbers $+\frac{1}{2}$ and $-\frac{1}{2}$. Other examples include the principal quantum number, $n$, which characterizes the energy levels of electrons in atoms, electric charge, quark colour charge, etc.

**Quark**. The elementary constituents of hadrons. All hadrons are composed of triplets of spin ½ quarks (baryons) or combinations of quarks and anti-quarks (mesons). The quarks form three generations, each with different flavours. The up- and down-quarks, with electric charges $+\frac{2}{3}$ and $-\frac{1}{3}$ and masses of 1.7–3.3 MeV and 4.1–5.8 MeV, respectively, form the first generation. Protons and neutrons are composed of up- and down-quarks. The second generation consists of the charm and strange-quarks, with electric charges $+\frac{2}{3}$ and $-\frac{1}{3}$ and masses of 1.27 GeV and 101 MeV, respectively. The third generation consists of bottom and top quarks, with electric charges $+\frac{2}{3}$ and $-\frac{1}{3}$ and masses of 4.19 GeV and 172 GeV, respectively. Quarks also carry colour charge, with each flavour of quark possessing red, green, or blue charges.

**Renormalization**. One consequence of introducing particles as the quanta of fields is that they may undergo self-interaction, i.e. they can interact with their own fields. This means that techniques, such as perturbation theory, used to solve the field equations tend to break down, as the self-interaction terms appear as infinite corrections. Renormalization was developed as a mathematical device used to eliminate these self-interaction terms, by redefining the parameters (such as mass and charge) of the field particles themselves.

**SLAC**. Acronym for Stanford Linear Accelerator Center, located in the Los Altos Hills near Stanford University in California.

**Special relativity**. Developed by Einstein in 1905, the special theory of relativity asserts that all motion is relative, and there is no unique or privileged frame of reference against which motion can be measured. All inertial frames of reference are equivalent – an observer stationary on earth should obtain the same results from the same set of physical measurements as an observer moving with uniform velocity in a spaceship. Out go classical notions of absolute space, time, absolute rest, and simultaneity. In formulating the theory, Einstein assumed that the speed of light in a vacuum represents an ultimate speed which cannot be exceeded. The theory is 'special' only in the sense that it does not account for accelerated motion; this is covered in Einstein's general theory of relativity.

**Spin**. All elementary particles exhibit a type of angular momentum called spin. Although the spin of the electron was initially interpreted in terms of electron 'self-rotation' (the electron spinning on its own axis, like a spinning top), spin is a relativistic phenomenon and has no counterpart in classical physics. Particles are characterized by their spin quantum numbers. Particles with half-integral spin quantum numbers are called fermions. Particles with integral spin quantum numbers are called bosons. Matter particles are fermions. Force particles are bosons.

**SSC**. Acronym for Superconducting Supercollider, an American project to build the world's largest particle accelerator at Waxahachie in Ellis County, Texas, capable of proton–proton collision energies of 40 TeV. The project was cancelled by Congress in October 1993.

**Standard Model, of big bang cosmology**. *See* lambda-CDM model.

**Standard Model, of particle physics**. The currently accepted theoretical model describing matter particles and the forces between them, with the exception of gravity. The Standard Model consists of a collection of quantum field theories with local SU(3) (colour force) and SU(2)×U(1) (weak nuclear force and electromagnetism) symmetries. The Model contains three generations of quarks and leptons, the photon, W, and Z particles, colour force gluons, and the Higgs boson.

**Strangeness**. Identified as a characteristic property of particles such as the neutral lambda, neutral and charged sigma and xi particles, and the kaons. Strangeness was used together with electric charge and isospin to classify particles according to the 'Eightfold Way' by Murray Gell-Mann and Yuval Ne'eman (see Figure 10, p. 69).

This property was subsequently traced to the presence in these composite particles of the strange-quark (see Figure 12, p. 82).

**Strange-quark**. A second generation quark with charge $-\frac{1}{3}$, spin ½ (fermion), and a mass of 101 MeV. The property of 'strangeness' was identified as a characteristic of a series of relatively low-energy (low-mass) particles discovered in the 1940s and 1950s by Murray Gell-Mann and independently by Kazuhiko Nishijima and Tadao Nakano. This property was subsequently traced by Gell-Mann and George Zweig to the presence in these composite particles of the strange-quark (see Figure 12, p. 82).

**Strong force**. The strong nuclear force, or colour force, binds quarks and gluons together inside hadrons and is described by quantum chromodynamics. The force that binds protons and neutrons together inside atomic nuclei (also referred to as the strong nuclear force) is thought to be a 'hang-over' of the colour force binding quarks inside the nucleons. *See* colour force.

**SU(2) symmetry group**. The special unitary group of transformations of two complex variables. Identified by Chen Ning Yang and Robert Mills as the symmetry group on which a quantum field theory of the strong nuclear force should be based, SU(2) was subsequently identified with the weak force and, when combined with the U(1) field theory of electromagnetism, forms the SU(2)× U(1) field theory of the electro-weak force.

**SU(3) symmetry group**. The special unitary group of transformation of three complex variables. Used by Murray Gell-Mann and Yuval Ne'eman as a global symmetry on which the 'Eightfold Way' was constructed. Subsequently used by Gell-Mann, Harald Fritzsch, and Heinrich Leutwyler as a local symmetry on which to base a quantum field theory of the strong nuclear (colour) force between quarks and gluons.

**Superconductivity**. Discovered by Heike Kamerlingh Onnes in 1911. When cooled below a certain critical temperature, certain crystalline materials lose all electrical resistance and become superconductors. An electric current will flow indefinitely in a superconducting wire flowing with no energy input. Superconductivity is a quantum-mechanical phenomenon explained using the BCS mechanism, named for John Bardeen, Leon Cooper, and John Schrieffer.

**Supersymmetry (SUSY)**. An alternative to the Standard Model of particle physics in which the asymmetry between matter particles (fermions) and force particles (bosons) is explained in terms of a broken supersymmetry. At high energies (for example, the kinds of energies that prevailed during the very early stages of the big bang) supersymmetry would be unbroken and there would be perfect symmetry between fermions and bosons. Aside from the asymmetry between fermions and bosons, the broken supersymmetry predicts a collection of massive super-partners with a spin different by ½. The supersymmetric partners of fermions are called sfermions. The partner of the electron is called the selectron; each quark is partnered by a corresponding squark. Likewise, for every boson there is a bosino. Supersymmetric partners of the photon, W and Z particles, and gluons are the photino, wino and zino, and gluinos. Supersymmetry resolves many of the problems with the Standard Model, but evidence for super-partners has not yet been found.

**Symmetry-breaking**. Spontaneous symmetry-breaking occurs whenever the lowest-energy state of a physical system has lower symmetry than higher-energy states. As the system loses energy and settles to its lowest-energy state, the symmetry spontaneously reduces, or 'breaks'. For example, a pencil perfectly balanced on its tip is symmetrical, but will topple over to give a more stable, lower-energy, less symmetrical state with the pencil lying along one specific direction.

**Synchrotron**. A type of particle accelerator in which the electric field used to accelerate the particles and the magnetic field used to circulate them in a ring are carefully synchronized with the particle beam.

**Tera**. A prefix denoting trillion. A tera electron volts (TeV) is a trillion electron volts, $10^{12}$ eV, or 1000 GeV.

**Top quark**. Also sometimes referred to as the 'truth' quark. A third-generation quark with charge $+\frac{2}{3}$, spin ½ (fermion), and a mass of 172 GeV. It was discovered at Fermilab in 1995.

**Trillion**. A thousand billion or a million million, $10^{12}$, or 1,000,000,000,000.

**U(1) symmetry group**. The unitary group of transformations of one complex variable. It is equivalent (the technical term is 'isomorphic') with the circle group, the multiplicative group of all complex numbers with absolute value of unity (in other words, the unit circle in the complex plane). It is also isomorphic with SO(2), a special orthogonal group which describes the symmetry transformation involved in rotating an object in two dimensions. In quantum electrodynamics, U(1) is identified with the phase symmetry of the electron wavefunction (see Figure 7, p. 34).

**Uncertainty principle**. Discovered by Werner Heisenberg in 1927. The uncertainty principle states that there is a fundamental limit to the precision with which it is possible to measure pairs of 'conjugate' observables, such as position and momentum, and energy and time. The principle can be traced to the fundamental duality of wave and particle behaviour in quantum objects.

**Vacuum expectation value**. In quantum theory, the magnitudes of observable quantities such as energy are given as the so-called expectation (or average) values of quantum-mechanical operators which correspond to the observables. The operators are mathematical functions which operate on, and change, the wavefunctions. The vacuum expectation value is the expectation value of the operator in a vacuum. Because of the shape of the potential energy curve of the Higgs field, it has a non-zero vacuum expectation value which breaks the symmetry of the electro-weak force – see Figure 13, p. 87.

**W, Z particles**. Elementary particles which carry the weak nuclear force. The W particles are spin 1 bosons with unit positive and negative electrical charge ($W^+$, $W^-$) and masses of 80 GeV. The $Z^0$ is an electrically neutral spin 1 boson with mass 91 GeV. The W and Z particles gain mass through the Higgs mechanism and can be thought of as 'heavy' photons.

**Wave–particle duality**. A fundamental property of all quantum particles, which exhibit both delocalized wave behaviour (such as diffraction and interference) and localized particle behaviour depending on the type of apparatus used to make measurements on them. First suggested as a property of matter particles such as electrons by Louis de Broglie in 1923.

**Wavefunction**. The mathematical description of matter particles such as electrons as 'matter waves' leads to equations characteristic of wave motion. Such wave equations feature a wavefunction whose amplitude and phase evolve in space and time. The wavefunctions of the electron in a hydrogen atom form characteristic three-dimensional patterns around the nucleus called orbitals. Wave mechanics – an expression of quantum mechanics in terms of matter waves – was first elucidated by Erwin Schrödinger in 1926.

**Weak neutral current**. A weak-force interaction involving the exchange of a virtual $Z^0$ boson or a combination of virtual $W^+$ and $W^-$ particles, see Figures 15 and 16, pp. 100 and 127.

**Weak nuclear force**. The weak force is so called because it is considerably weaker than both the strong and electromagnetic forces, in strength and range. The weak force affects both quarks and leptons and weak-force interactions can change quark and lepton flavour, for example turning an up-quark into a down-quark and an electron into an electron neutrino. The weak force was originally identified as a fundamental force from studies of beta-radioactive decay. Carriers of the weak force are the W and Z particles. The weak force was combined with electromagnetism in the SU(2)×U(1) quantum field theory of the electro-weak force by Steven Weinberg and Abdus Salam in 1967–68.

**Yang–Mills field theory**. A form of quantum field theory based on gauge invariance developed in 1954 by Chen Ning Yang and Robert Mills. Yang–Mills field theory underpins all the components of the current Standard Model of particle physics.

# BIBLIOGRAPHY


Baggott, Jim, *Beyond Measure: Modern Physics, Philosophy and the Meaning of Quantum Theory*, Oxford University Press, 2003.

Baggott, Jim, *The Quantum Story: A History in 40 Moments*, Oxford University Press, 2011.

Cashmore, Roger, Maiani, Luciano, and Revol, Jean-Pierre (eds.), *Prestigious Discoveries at CERN*, Springer, Berlin, 2004.

Crease, Robert P. and Mann, Charles C., *The Second Creation: Makers of the Revolution in Twentieth-Century Physics*, Rutgers University Press, 1986.

Dodd, J.E., *The Ideas of Particle Physics*, Cambridge University Press, 1984.

Enz, Charles P., *No Time to be Brief: a Scientific Biography of Wolfgang Pauli*, Oxford University Press, 2002.

Evans, Lyndon (ed.), *The Large Madron Collider: A Marvel of Technology*, CRC Press London, 2009.

Farmelo, Graham (ed.), *It Must be Beautiful: Great Equations of Modern Science*, Granta Books, London, 2002.

Feynman, Richard P., *QED: The Strange Theory of Light and Matter*, Penguin, London, 1985.

Gell-Mann, Murray, *The Quark and the Jaguar*, Little, Brown & Co., London, 1994.

Gleick, James, *Genius: Richard Feynman and Modern Physics,* Little, Brown & Co., London, 1992.

Greene, Brian, *The Elegant Universe: Superstrings, Hidden Dimensions and the Quest for the Ultimate Theory*, Vintage Books, London, 2000.

Greene, Brian, *The Fabric of the Cosmos: Space, Time and the Texture of Reality,* Allen Lane, London, 2004.

Gribbin, John, *Q is for Quantum: Particle Physics from A to Z*, Weidenfeld & Nicholson, London, 1998.

Guth, Alan H., *The Inflationary Universe: The Quest for a New Theory of Cosmic Origins*, Vintage, London, 1998.

Halpern, Paul, *Collider: The Search for the World's Smallest Particles*, John Wiley, New Jersey, 2009.

Hoddeson, Lillian, Brown, Laurie, Riordan, Michael, and Dresden, Max, *The Rise of the Standard Model: Particle Physics in the 1960s and 1970s*, Cambridge University Press, 1997.

Johnson, George, *Strange Beauty: Murray Gell-Mann and the Revolution in Twentieth-Century Physics*, Vintage, London, 2001.

Kane, Gordon, *Supersymmetry: Unveiling the Ultimate Laws of the Universe*, Perseus Books, Cambridge, MA, 2000.

Kragh, Helge, *Quantum Generations: A History of Physics in the Twentieth Century*, Princeton University Press, 1999.

Lederman, Leon (with Dick Teresi), *The God Particle: If the Universe is the Answer, What is the Question?*, Bantam Press, London, 1993.

Mehra, Jagdish, *The Beat of a Different Drum: The Life and Science of Richard Feynman*, Oxford University Press, 1994.

Nambu, Yoichiro, *Quarks*, World Scientific, Singapore, 1981.

Pais, Abraham, *Subtle is the Lord: The Science and the Life of Albert Einstein*, Oxford University Press, 1982.

Pais, Abraham, *Inward Bound: Of Matter and Forces in the Physical World*, Oxford University Press, 1986.

Pickering, Andrew, *Constructing Quarks: A Sociological History of Particle Physics*, University of Chicago Press, 1984.

Riordan, Michael, *The Hunting of the Quark: A True Story of Modern Physics*, Simon & Shuster, New York, 1987.

Sambursky, S., *The Physical World of the Greeks*, 2nd Edition, Routledge & Kegan Paul, London, 1963

Sample, Ian, *Massive: The Hunt for the God Particle*, Virgin Books, London, 2010.

Schweber, Silvan S., *QED and the Men Who Made It: Dyson, Feynman, Schwinger, Tomonaga*, Princeton University Press, 1994.

Stachel, John (ed.), *Einstein's Miraculous Year: Five Papers that Changed the Face of Physics*, Princeton University Press, 2005.

't Hooft, Gerard, *In Search of the Ultimate Building Blocks*, Cambridge University Press, 1997.

Veltman, Martinus, *Facts and Mysteries in Elementary Particle Physics*, World Scientific, London, 2003.

Weinberg, Steven, *Dreams of a Final Theory: The Search for the Fundamental Laws of Nature*, Vintage, London, 1993.

Weyl, Hermann, *Symmetry*, Princeton University Press, 1952.

Wilczek, Frank, *The Lightness of Being: Big Questions, Real Answers*, Allen Lane, London, 2009

Woit, Peter, *Not Even Wrong*, Vintage Books, London, 2007.

Zee, A., *Fearful Symmetry: The Search for Beauty in Modern Physics*, Princeton University Press, 2007 (first published 1986).

# INDEX